TEXAS

THE PLACE TO LIVE

T. F. BUCK

edited, with new material by Michelle M. Haas

Copano Bay Press
2013

Originally published in 1860 by Mendenhall (Cincinnati), under the title *Texas: The Australia of America; or The Place to Live* with an anonymous byline of "A Six Years' Resident."

Copano Bay Press

ISBN: 978-0-9884357-6-6

Contents

Publisher's Note

This book is presented in memory of Talcott Frank Buck, an early Texas cattle driver who wanted to better the world.

-Michelle M. Haas, Managing Editor
Windy Hill

1867 depiction of cattle branding on Texas prairies. Taken from a sketch by James. A. Taylor, as published in *Leslie's Illustrated Newspaper*.

Branding a Maverick: An Introduction to Talcott Frank Buck

by Michelle M. Haas

It's pretty tough to do anything anonymously in the 21st century. Everything from our email correspondence to what we buy at the grocery store is tracked by someone or other. It was a little easier to fly under the radar in 1860, though, and many a man published a book without affixing his name to it. Time and technology have flushed some of these men out of the shadows of anonymity, but others, like the author of this book, have stood quietly by. Before I raise the veil on the man, I should tell you a little bit about how he was coerced out of hiding.

During the early phase of the book's development, it nagged me that the author only billed himself as "A Six Years' Resident." Sure, that added a certain air of mystery to the book but it didn't satisfy my curiosity. Who was this guy? Where did he come from and how did he end up in Texas? Did he leave? Where did he go? When did he die and how?

He seemed intentionally vague about his place of origin, except to say that he was not from New England. I tried to take clues from how and what he wrote. The text tells us that he visited several countries controlled by the British crown. Some rogue peculiarities of spelling made me wonder if he might be Canadian. But that theory didn't pass the smell test. Some spellings that needed to be there to prove it out were absent. There was very little else there to pick at. Back to good ol' square one. Even when submitting his book to the Library of Congress many years after its publication, he didn't include his name to be recorded.

The book's publication could not have been more ill-timed. The Civil War was in the making. The book took on quite an anti-abolitionist hue. The publisher was located in Cincinnati, which was at the forefront of the abolitionist movement. Whatever his personal stance on slavery, the publisher would not have been best served by promoting a book singing the praises of agricultural opportunities in a southern state.

In 1860, at least two people knew who wrote the book: the author and his publisher. And ultimately it was his publisher, Edward Mendenhall, who gave us our man. In 1868, an enormous wholesale catalogue intended for the book trade listed all of Mendenhall's publications then for sale. Among them was this book, but the author's name was kept anonymous. The wholesale price to booksellers was 34 cents, by the way. Retail cover price was 50 cents.

Hop to 1881, when Frederick Leypoldt, noted publisher, bibliographer and founder of *Publisher's Weekly*, issued his first complete *American Catalogue*. This valuable work presented all books for sale from hundreds of American publishers through July 1, 1876. It took Leypoldt another five years to organize and prepare the book for publication. Mendenhall was among the publishers included and, this time, whether through a clerical goof or intention, he attached a name to the book's bibliographic record: *T. F. Buck*.

A full name would have been a boon, but I'm not one to look a gift horse in the mouth. A quick search on the name in databases and newspaper archives returned some promising results. After a lot of parsing of information, I eventually found proof enough to confidently attach Mr. Buck's name to this book. I found it in his dying words. His life was a troubled one.

One T. F. Buck, who hailed from Lowville, Lewis County, New York (about 100 miles from the Canadian border), seemed to be the most likely candidate. Born about 1825 to Daniel Talcott and Laura White Buck, the man's full name was Talcott Frank Buck. He was of Revolutionary War stock and his family led a typical American agricultural life for generations. In 1850, we find Talcott in Wayne County, New York, boarding with a jeweler, perhaps as an apprentice. He married the girl next door (or as close to the girl next door as you can get in rural 1850s New York State), Miss Margaret J. Mandeville in 1851. Their first child, Daniel Mandeville Buck, was born in September of 1852 in Wayne County. At this point in Talcott's timeline, the events that would lead him to Texas began to form up and start marching.

About the time he married Margaret, Australia's gold rush was underway and the tongues of the press were wagging all over the world. Tons of gold were being shipped daily from Australia to the motherland! Fed up with the rat race and being poor in a world where new ways to get rich quick were a dime a dozen, Talcott set out for Australia, probably before the birth of his son. He may not have even known yet that he was going to be a father. His journey entailed a pit stop in

South Africa where he spoke to the locals about abolition and slavery, then on to Australia where he tried his hand at mining and queried the locals about ranching in their fair climate.

Short on funds and not having made his fortune in the gold mines, he soon struck back for the States. This time his journey sent him in the other direction, with a stop in Central America. At the time, Nicaragua was in the midst of an on-again, off-again civil war. The Mosquito Coast was enmeshed in controversy over who controlled it and it was from the port of San Juan del Norte that Buck sailed for New Orleans, arriving there in March of 1853 aboard the Vanderbilt-owned side wheeler, *Daniel Webster*. Factoring in his proximity to Texas upon landing, his lack of funds, the distance to his wife and child in New York and the timing of the birth of his next kid, it's a safe bet that Talcott Buck made a beeline for Texas from New Orleans. With no fortune obtained down under, he needed a new start.

Buck describes his occupation while in Texas as that of a purchaser and shipper of livestock for a wealthy company. His knowledge of the Goliad area and the birth of his second son in Lavaca County point to employment with James M. Foster, the man who pioneered the New Orleans market for Texas cattle. Foster had exclusive access to the Morgan lines out of Indianola for livestock shipping. He employed agents, such as Buck, to purchase cattle and round up herds for shipment. During his time as a cattle driver and shipper, Buck spent time with Thomas O'Connor, Thomas Decrow, the Fagans and other pioneering South Texas ranchers. He no doubt bumped into a young and green

Shanghai Pierce during visits to the Grimes Ranch and may have encountered Kenedy and King during his sojourns to the Nueces Strip after cattle.

The text touts the virtues of Texas as the most temperate, beautiful and wide-open place on the planet to live. It then spins into a diatribe about how poor folks from the congested, cold North should come settle up the land in Texas, since Texans, for the most part, weren't doing much to improve it—the free and open ranges made it too easy for a man to earn a living, says Buck, and this made many Texans lazy. They weren't doing the most they could with what they had, he thought.

I took umbrage to this attitude. Obviously this man knew little about our history or the stout people who carried Texas from the empresario days through the Revolution, the Republic and on to statehood. Living in early Texas wasn't exactly a cake walk. It was a hard life that you had to put your back into. I immediately questioned: Where's *your* fortune, buddy? If you're such a hard working man, where's your ranch, your herd, your land after six years? I asked and Talcott Buck answered me. He tried to get rich quick and fell flat on his face. (My guess is that he lost his shirt in some kind of land speculation deal.) He readily admitted his mistake and wanted this book to be a cautionary tale for others. Do as Buck says, not as Buck did. Anyone with a little pluck could make it big in Texas, said he. Here, a man can be a man and not a dog.

Talcott Buck managed to get his wife, Margaret, and their youngest son, Daniel, down to Texas by early 1856. Whether father and son had met prior to this, we have no way to know. In October of 1857, another son, Walter White Buck, was born to the couple in Lavaca County. Buck writes fondly of his oldest child feeding the cardinals by their front door and of plans to take his little boy with him to herd the flock of sheep he would one day own in Texas.

He quit the livestock business, though, to write this book. And if you doubt for one minute the truth of that statement, think again. As we will soon see, this book was only the beginning of a body of work that was intended to serve a far greater purpose than just encouraging emigration to Texas.

By the time of the book's publication, Talcott Buck's younger brother, Ebenezer, had taken up residence in Texas, too. He was laboring as a hay cutter in Calhoun County in the summer of 1860. An older brother, Ralph, owned a farm in Ohio and it was there that Talcott Buck headed next, probably to pick up and distribute some of the books he had printed in Cincinnati in March. En route from Texas to Cincinnati, probably by way of New Orleans, Mrs. Buck came down with a fever. Twenty miles beyond Ralph's farm, Margaret J. Buck died, near Lordstown on May 15, 1860. She was buried in Columbus.

Talcott hung around Ohio for a few years. The Civil War was a reality and he was a single father of two boys. His brother Ralph was widowed in 1861 and the two brothers were remarried within 2 days of each other in mid-September 1863 in different Ohio towns. Talcott married a divorcee, Mrs. Elsie PerLee Jarvis, a New York native who was roughly his age. When the Civil War dust cleared, they made their way back to Texas.

In February 1867, she bore him a son here. Reconstruction-era Texas doesn't seem to have impressed either New Yorker and the Bucks departed for points north. A girl child was born to them in New York in 1869.

1870 finds Elsie Buck back at her family home in Lenox, New York with their two children. Talcott has gathered up his kids from his first marriage, and he and his eldest son Daniel are driving cattle together in Missouri.

Talcott's brother, Ebenezer, had fought for the Confederacy in the war, enlisting at Calhoun County. He had established a farm in Refugio County after the war and established a family. In 1873, he moved operations from Refugio to Pipe Creek in Bandera County. In the same year, Talcott returned to Texas and was appointed postmaster of Blooming Grove, near Corsicana. He finally mailed a copy of his book, published thirteen years earlier, to the Library of Congress to be catalogued.

In 1876, the fate of the Buck clan takes a dark turn. Whether it was the strife of the Civil War and Reconstruction, or perhaps he was just born out of sorts, Talcott Frank Buck began to come undone. He had seen his nation divided, despite having made suggestions in this very book on how that conflict may have been prevented. He had never enjoyed the simple pleasure of herding sheep through Texas with Daniel. Texas and the world were changing and Talcott couldn't keep up. As daily life, family business and change moved all around him, he was stuck in an endless loop of misery and looking for a solution.

Foreshadowing of what was to come can be seen in this very book. Something was tugging at his sleeve in 1859-60 as he was writing it. It's hard to miss if you know what to look for. The themes of mankind's betterment through emigration, and the purity of nature are present in not a subtle way. To many of us, it may read like the protests of an aging man, squawking about kids these days with their laziness and bad manners, bad weather, and politicians being crooks...the rich keep getting richer, while the poor work harder but get nowhere. But for Talcott Buck these things weren't mere observations on the state of the world—they were problems to be solved and the burden rested squarely on him. He loathed the corruption and deceit he read about in the papers. He came to loathe the idea of people practicing Christianity for the sake of the afterlife rather than for the sake of leading the pure, wholesome lives that God intended us to live on earth. Talcott believed that he had the power to set the world right, to make the government and mankind see the error of their collective ways. He believed he could move the great big world with his writings and ideas. His book on Texas was his first gentle attempt. His next would be less gentle.

In 1876, during the Centennial Exposition at Philadelphia, American democracy celebrated one hundred years of life. On display were conspicuous shows of patriotism, alongside world-altering inventions like the typewriter, the telephone and the precursor to the electric light. All things new and exciting were abuzz at the Exposition. The future was there. Talcott was there. He may have made a scene or perhaps attempted suicide, and was arrested as an insane person. At the asylum, the white shirt he wore was marked inside with his last name and the initials "M.D."—mentally deranged. He would wear the shirt again later under similar circumstances.

Buck had been writing letters to newspapers, bankers, prominent orators, governors, U.S. presidents and congressmen, trying to get the ball rolling on the purification of the human race on earth. He cited the agnostic and freethought philosophies of Robert Ingersoll and the Darwinism of Henry Ward Beecher. But nobody was listening. The 1876 episode was probably his first non-written attempt at getting the world's attention so that he might rescue it.

When 1880 rolled around, his second wife was caring for their two children alone in New York, had obtained a divorce and had dropped the Buck name. Daniel and Walter, Talcott's boys from his first marriage, were back in Texas. The eldest Buck boy was ranching near Corsicana. Young Walter was working as a jeweler in the same area. Their uncle, Ebenezer, remained in Bandera County. The Texas roots planted by Talcott before his mind went south had staying power.

By 1883, Buck had hatched his master plan to fix mankind. If the powers that be wouldn't get on board and help him, he'd just right the ship himself. After staying with his brother in Bandera County for a time, giving speeches and writing letters to the paper about the ills of mankind, he headed north. Lodging at the home of a widowed Irishman, Lewis Kavanagh, Talcott spoke to his kindly landlord about what a wicked world they were living in and how he intended to reform it single-handedly. He spent most days in his room pouring words onto pages that would total several thousand by the time of his death. Buck never talked about what line of work he was in, but he made fond mention of the time he spent working livestock in Texas.

Buck left the Irishman's house in early 1884 and took up residence in room 148 at the Astor House in New York City. There, on January 8, Buck took some poison, but the poison didn't take. He made a scene. The brief coverage given the incident by the New York papers indicate that he violently rang the bell at the desk. When a hotel employee emerged, he behaved so erratically that a doctor was called. He was declared to be insane and hauled down the police station, then admitted to Bellevue Hospital's newly opened psych ward. From there was a transfer to the notorious Flatbush Insane Asylum in Brooklyn. After Buck's release from the institution, whether because they'd "cured" him or because of overcrowding, he was sent home to his sons in Navarro County, Texas.

It took him many months to scrape together the funds to take another stab at martyrdom and he was becoming increasingly paranoid that the press and others were trying to thwart his efforts. He deposited the bulk of his writings upon religion and the state of mankind with his son Walter, instructing him to place them in a safe deposit box at the bank in Hillsboro. He gave another box of his writings—the ones he had intended to be found with his dead body at the Astor House—to his son Daniel. Then he embarked on his final journey.

He wended his way to Washington, D. C. in early August, 1885, in hopes of a chance meeting with Robert Ingersoll. Buck thought that once Ingersoll knew his purpose, the great orator would put him up in a hotel where he would write a grand speech intended for President Cleveland and Congress on purification and life. But he did not bump into Bob Ingersoll. So he decided to go with Plan B.

Setting the date of his suicide as the date of Ulysses Grant's funeral, August 8, 1885, he put up at the Parker Hotel and purchased vials of morphine and laudanum (opium dissolved in alcohol). He prepared letters to newspapers from Texas to New York, as well as governors and former governors of Texas. He wrote to Henry Ward Beecher and Robert Ingersoll and advised them that he was entrusting them with the publication of his "works" after his demise. "If you would purify the race and lift this wicked world from its immensity of needless suffering, be sure that this mass of stuff in the hands of my boys is published... There never was and never will be another such struggle. But the world is sure to find out that I am right, and that must be my recompense." He beseeched Henry Ward Beecher, in somewhat ironic language, "In Heaven's name, read my works and find yourself, and then do something for this world before you die."

He presumed that when these letters were found with his corpse, they would be forwarded to their intended recipients. Then the world would have to sit up and take notice. But even at the height of his delusional journey, Talcott Buck was conscious of the fact the funeral of U. S. Grant, with its seven mile stretch of mourners, would probably dwarf his suicide in the papers. It's tough to share the limelight with a dead president, so he called it off and rethought things.

He headed back to the friendly home of Lewis Kavanagh in Jersey City on September 3 and inquired about the rental of a room. Kavanagh had no rooms that evening, so Talcott slept at the Wagner Hotel. The next day, he bunked at Kavanagh's and they spoke about a Hoboken pharmacist who had made headlines recently by committing suicide. Buck said that it was indeed foolish for a man to take his own life. The next morning, Saturday, Buck said he was going to visit with some relations on the Hudson River and that was the last Kavanagh heard from him until he received a peculiar package in the mail on Tuesday. Buck had written a letter to Kavanagh on the night he lodged with him. When he left the Irishman's home, he walked across the street to the Eagle Hotel, bummed an envelope and dropped his missive in the mail.

From Jersey City, he crossed into the Big Apple and checked in to the Home Made Hotel. His plan to get noticed on that day was to walk into a newspaper office, plop his writings down on a desk, then blow his brains out. But he changed his mind and wrote at the hotel. On Sunday morning, he went out to Central Park to scout a suitable place to "finish the mission," but he found all of the secluded spots occupied by young people enjoying the weekend. He slept at the hotel again Sunday evening and checked out on Monday morning, paying this bill and telling the clerk he was headed back to Washington. The truth, though, was that he was out of money, so it was back to Central Park.

He bought a newspaper on the morning of Monday, September 7, on which he made notes about the Park. "Central Park is beautiful. Were it not selfish [of] me,

I should like to live here a hundred or two years. But there are too many suicides and too much misery in the world. I must overcome it." He also made a note next to it saying that although there was a shot in the locker (a bullet waiting in the gun), it was hard to die.

Throughout the day of the 7th, he wrote on. To former president Chester Arthur he wrote: "I wrote you in the greatest distress that I was going to help the world. You paid no attention to me. What have you now to say about the purification of the race?" Letters to Texas Governors Lubbock, Roberts, Ireland and Hubbard were of the same bent.

Buck had given a lecture on religion, corruption and purity in Bandera and had written at length to the local newspaper during his stay with his brother in that town. From Central Park, he wrote to the editor of the *Bandera Bugle*:

> Doubtless you have long thought me dead. If not in person, at least so in the cause of reform. But not at all dead. I am still pulling away in the most matchless of courses, and although I failed in my first effort to strike a perfecting light, I am soon to try again, I hope with better success. They say the world moves. I wouldn't wonder if it did, and it seems to me that I am going to be a sort of world-mover. I wonder what you will have to say about it, if anything.

On another sheet of paper, labeled "Second Very Last Words," he wrote on the 7th:

> Last evening I accused myself of criminal delay and promised myself I'd surely go this morning. I

> got my breakfast this morning and started to push things right through, but here I am, after looking at the animals, with one excuse or another and still not dead. I must say that I am astonished to find the monkey or chimpanzee so much more like a human being than I had supposed. Well, that chimpanzee is a link in the chain of human species. I believe, as Beecher says, the day may come when they will find all the connecting links. This very animal proves to me that man is off the track, overrating himself—in a chimerical sense—that is, in a sense of nothingness, and underrating himself in reality.
>
> I have just counted my money and have fifty-nine cents. I cannot say that I am either glad or sorry. Had I more, I might find an excuse to go over another night. It is evident my time has come. If I were offered Vanderbilt's palace at this moment, I would not take it unless he would do as I am now going to do.

And finally, he prepared individual envelopes to be sent to newspapers from San Antonio to his hometown of Lowville, New York and most big city newspapers. He bundled them (each individually addressed and postage paid) into a paper-wrapped package and upon the wrapper he wrote instructions to the finder of the letters, to open them or mail them just as they were. He attached a sheet to this massive clutch of letters that said:

> To all the newspapers whose names I would recall if I could—You must help me take this naturally

> pretty and perfect world out of the artificial hell in which it is so fearfully yet needlessly engulfed. Fear not—the people will go with you. Fear not—I will take upon myself all the hell fire punishment there is for any and all of you in the great bugbear, nowhere hereafter.
>
> I have given many years of my life and all I have earned, and now I give my life to the world. No other man has ever suffered the amount of misfortune that I have endured from fear of failure in my great work, in rescuing the race from its needless corruption and woe...
>
> To the editors of New York and the world—dare not refuse the publication of my writings. In this suicide and my manuscript works lie a million times the dearest birthright of man—that is purity—and to all intents and purposes perfection of universal life.

Three final thoughts came to Talcott Buck as dawn rose on the morning of the 8th of September 1885 in Central Park. First, he needed a summation of his beliefs—to clearly express what a better place the world would be if not for the corruption in religion and belief in creation and an afterlife. He took out a piece of paper and in what the *New York Daily Graphic* described as "odd, fat letters" wrote the following:

> The chit of the world's redemption lies in a universally established fact of No Life After Death, which I shall have sooner or later established through the earth. This fact was established on the creation of

things by the Almighty God. But cowardly, impious man for some reason is overriding and trampling underfoot. And that is one grand reason why the world cannot get pure and perfect.

After failed attempts and stalls in his bid for the salvation of humanity, he wanted to be sure that he actually died once he had committed the deed, so he scribbled down some last words:

> These are the very last words I intend ever to write. Sorry I am without a change of clothes. I put these on in Washington, thinking this job would be over within a day or two after leaving there. What is the difference, providing I help the world! My good people, my good doctors, do no try to save me in case the bullet does not finish me. Rather, give me opiates and let the poison within me do its work. The world will be the gainer and I will be free of pain. Goodbye, world. Be good to yourselves and pacify. —T. F. Buck

Among his last ramblings in the wee hours of the morning of September 8, 1885, he wrote of Texas, and removed all doubt that I had found the author of this long-forgotten Texas ranching book:

> Had I a million dollars, or a million and a thousand dollars, I would give the million to anyone who would take my place and do this job for me. Yes, I would give the million and $900 of it. With the $100 left me, I'd get back to Texas where I'd go to work and soon have me a nice little home. Yes, by the eternal! I'd give the last cent of it and beg my

> way back to Texas, and as old as I am, I'd make the desert shine for a few years at least. But don't let me fool anybody. It isn't everyone who can make ends meet so nicely in Texas nowadays. But it is a good country for those who will, &c. Men are somewhat constituted for peculiar climate and peculiarities of particular sections of the world, and thrice happy will be the immigrant who strikes the country best adapted to his constitution.

Talcott Frank Buck then placed all of his letters in his brown satchel—the only luggage he had brought with him from Texas. He fixed up a rope out of two handkerchiefs tied together and bound himself to the bench he was occupying in a Central Park pavilion, so that he wouldn't fall to the ground and become separated from his satchel full of writing. Swallowing enough of the laudanum from the vial to kill him outright, he had halfway completed his mission. He then took a .45 calibre Smith & Wesson and launched a bullet into his right temple, creating a wound which one brutal New York paper described as one "through which a baby might have put its tiny fist." Officer Patrick Meehan was at the Sixth Avenue entrance, near the pavilion, when he heard the shot and rushed to the scene. He found Talcott Buck there on the bench, held upright by the makeshift rope, his face covered in blood. His skull was shattered and his stomach contained enough poison to kill a horse. He gasped twice and then his mission was complete.

Talcott and his valise were taken to the police station at the Arsenal to await the coroner's arrival 4 hours later. The coroner concluded that the amount of laudanum he had ingested would have killed him rather quickly, had the bullet not done the job even sooner. Buck wore a long, grey beard. He was dressed as though he were still in Texas in 1860—square-toe boots, his white shirt (from the 1876 Exposition incident), a blue flannel coat, dark blue velvet vest, gray trousers and an old grey hat. His satchel was searched to learn his identity and next of kin.

The letter that Buck had mailed on Saturday to Lewis Kavanagh in Jersey City arrived on the morning of his death. Its contents bewildered the Irishman. Included was a diary of sorts that Buck had started while in Washington, as well as some other papers. He instructed Kavanagh to sell them for ten or fifteen thousand dollars. Then he was to take some of the money to better his situation.

"Write to my two boys," Buck said, "and turn their share of the proceeds over to them and they will see that my brothers and sisters are helped, and if they are dead and have left needy children, I want them helped." He further included a note to Daniel and Walter, telling them not to disregard the wishes of the dying and to use the money from the sale of his important papers to take care of his kin. He closes by fretting that he's down to his last $5.50, which won't buy him enough time to get prepared, so he concludes he'll have to be dead by September 5th.

From the letter, Kavanagh still wasn't quite clear what had happened or was happening to Buck. The newspaper reporter conducting the interview informed him of Buck's suicide and he was "very much affected" by the news. He described Talcott Buck as an honorable man and a gentleman.

The newspapers in New York City carried the story because suicide sold papers. Odd suicides sold more papers. The headlines were, of course, mocking and sensational in tone:

One Less Texas Crank

Crazy Texan Offers Himself Up

Another Park Suicide

TF Buck Blows His Brains Out in Central Park

Several papers gave one full column to the story because it was so weird. Were it not for his notes and the newspapers' coverage of their contents, I would be without the material necessary to write this introduction. No mention was made of where he was to be buried. Suffice it to say, the former presidents and governors and newspapers did not receive, acknowledge or act on his prolific writings. The world is no less corrupt and no more pure because he ended his life. He was a martyr in his own mind alone. His quest to improve the world began, though, with this book about Texas and the time he spent here was the only time that he was at peace, it seems.

In addition to insights on the earliest days of commercial ranching in Texas, Talcott Frank Buck gave us the gifts of his two sons, Daniel and Walter Buck. Daniel, who had driven cattle with his father as a young man, began a life of farming in Corsicana but removed to Bosque County to become a jeweler and raise a family. He died there in 1936.

Walter Buck, who started out as a jeweler in Corsicana, worked the land in Hill County and ended up a stockraiser in Junction, Texas in 1910. His son, Walter W. Buck, Jr. followed the family tradition, inheriting the land and adding acreage when his father died in 1932. He raised cattle and goats there until he was ready to part with it. In 1977, he donated 2,200 acres to the State of Texas to form the Walter Buck Wildlife Management Area. An adjacent 500 or so now form the South Llano River State Park, one of the most beautiful in Texas. It is said that Buck donated the land so that one day city folks could go up on the hills that he worked every day and experience the beautiful sights he got to enjoy.

Talcott Frank Buck may not have made a fortune, saved the world from itself or sold many books. The human race is still the human race with all its foibles. But his books are being sold now. His story is being told now. Texas knows who he is now. It may not be the recognition he thought would save the world, but from where I sit, in relative proximity to what was his 1850s South Texas utopia, I say it beats the hell out of another century of obscurity.

Author's Preface

To drop the plow, spade, ax, and the like, and travel over the earth until a country is found where these companions of the laboring man need not swing so heavily, where his task need not be so hard, when his condition in life may be easily bettered, and then write home and to all the world, in the shape of a book, and tell poor mankind of such discovery, is what men seldom do. Men may explore the remotest ends of the earth, and draw the heavens near for geographical, historical, and philosophical purposes, and they may crowd the world with their histories and important discoveries. They may write of love, of daring deeds and lofty purposes, and upon almost all the subjects that commonly suggest themselves to the mind. They may peruse the Holy Bible even, and write theological works enough to convert the whole world to Christianity; and yet the poor man may look in vain among all these writings to find how to get his bread and raiment, how the means to educate his children, or how his pecuniary wants are to be supplied, and his earthly career made more pleasant to himself and family.

The world is full of book producers, but how few address themselves to the legions who are anxious to know where they can go, what they can do, and how they are to do it.

The following pages were not designed so much for the literary world as for the generality of men. Nor were they designed so much to delight as to benefit. They were written principally for those who are and may be looking abroad over the earth for another and better range of operation, or a more accessible grainery of life; and although by the "huge paw" of a working man, the interested reader will not stop to criticize, but will leniently consider that the writer is one of those poor but somewhat observing laborers, who has snatched a moment from time to tell the world of a field that only awaits the sickle to reap its golden harvest.

As I have said, these pages may be considered as having been written to benefit rather than to delight. If the object is attained, what matters it to the reader that the writer is a bungler or anything but an accomplished composer? So, if the reader is of those for whom this work is written, it is hoped that he may not stop to notice faults and errors, but read on and see if there is anything here suggested that may be to his advantage.

A portion of this work was written more than a year since, which will account for any seeming conflicting statements in regard to time.

INTRODUCTORY

Kind reader, take, if you please, an imaginary flight over the densely populated portions of the civilized world, and consider how many there are who would do well to emigrate.

Ask, in your flight, the legions with whom you meet if they would wish to emigrate, and then calculate the number whose answer was "yes, but where shall I go?" No doubt such an answer, or one to the same effect, would fall from the lips of many thousands; and to many of those whose wish it is to find another and better country, the contents of this book may be somewhat interesting.

How often has the poor discontented man retraced the fruitless efforts of his past, looked around upon his unhappy present and, gazing out upon the dark uncertain future, exclaimed with feelings of anguish, akin to despair, "O, what can I do, or where shall I go? Have I left anything undone that I could do in this unpromising land? Must I remain forever here, penniless and destitute? Is there no spot upon this widespread earth whereon I could live and accumulate a little to be called my own?"

Is the reader a poor man of family, whose house may be open and almost roofless, and whose children are poorly clad, and the cold blasting snowstorms of another winter coming on, and do these shivering children stand around and cry aloud for a father's care?

Well might the father say, "This is too much; but O, God, what can I do? Where, O, where shall I go?"

Perhaps the reader is a man of property, and having his credit too far extended, it would now take one-half he has got to answer the demands against him, and he may say, "I am already as deeply in debt as my merchant will suffer me to go. There is no hope of making anything here; and if I were to make a desperate effort to dispose of my effects, settle up my business, and start out in the pursuit of a more promising country, which my little balance might accumulate, where my efforts might be crowned with success, where I might live free from debt and the angry demands of creditors, where I might make life more an object and live out a peaceful decline of years, and give my children a liberal and respectable start in the world—where," he may ask, "shall I go to find that country?"

Should chance open my book to the perusal of any whose feelings or situations might answer such or a similar portrayal, then let them read on. Methinks the reader is a young man whose cradle of infancy was no stranger to wealth; whose boyish days were happy in the sportive play of schoolmates; whose youthful mind was ever elate with the brightest of prospects, and the green meadows of his father's land bade him smile, be joyous, and hopeful; but by the vicissitudes of fortune he is now rendered poor and unhappy, and would fly away from the taunts and jeers of those who were once proud to be considered his friends and equals.

It may be you are an orphan boy, whose ears are unaccustomed to words of kindness, whose last blessing fall from the lips of a dying parent and, galling in the chains of discontent, would get away if you but knew where to go. If so, cease the mingling of tears with thy sorrows, and read on.

Possibly you are the resident of some densely populated country, where every trade, occupation, and pursuit is overdone, and where you cannot continue to live with hope of success in any business that you may be qualified to carry on.

Are you the daring sailor, for whom the perils and hardships of the sea have no further charms and, having furled the sail and bid adieu to hoarse commands, would you now seek that peace and profit which the billows of the ocean have failed to give? Have the prayers of an anxious and loving wife, whose mind has already too long dwelt upon the watery graves of the deep, induced you to remain upon land?

Perhaps you are the friend of someone whose wish it is to leave behind the land of his nativity and look for wealth, honors, or contentment in another. If so, please read on; and should you deem it worth your while, call the attention of your friend to this book.

Are you the laboring man of intellect, whose only want to render you worthy of your immortal nature, is leisure? Would you burst the fetters of poverty while yet in the vigor of life, in anticipation of declining years, and accumulate a little to smooth the roughness of cares in your old and feeble days?

Are you the emigrant from eastern or European shores, now landing upon American soil, and would you find a beautiful climate and most excellent locality for your future abode in this great land of liberty? Or are you the farmer's son, whose minor days are quite or nearly out, and would you now mingle with the tide of emigration, taking with you your habits of industry, to some distant country to work out your destiny of weal or woe? Be you who you may, if your future is unpromising—if you have carefully looked around you and can nowhere discover anything that merits attention, or, in other words, if you would better your fortune, read on.

Be you who you may—Briton, German, Frenchman, American, or what not—whether living upon the snow-clad hills of New England, the lake-washed shores of the Canadas, the heights of the Highlands, or in the valley of the Rhine—if you are discontented with your present abode, and are determined to seek a fortune or better your condition in another part of the world, then patiently read on. It may be that some of you will be interested. It may be that some of you will be benefited.

I would not create an unnecessary excitement; I would not lead one single human being astray; I would not induce the rugged son to leave the aged and helpless parent to mourn the departure of his undutiful child; I would not have anyone believe that anywhere upon the face of this widespread earth there is a country the spontaneous growth of which renders the labor of man unnecessary; but I would have thousands and tens of thousands of human beings believe that if they were in a country with which I am acquainted and about which I am going to write, they would be by far better off than they now are or can ever expect to be by remaining where they are.

I have no newly discovered gold or silver mines, no second California or Australian gold excitement upon which to enlarge, but I have a country, dear reader, of which to tell you, where, if some men could be, the prattling of their babes and the music of birds would be sweeter to their ears; where the frown upon their brows would involuntarily dispel, and the tears of their children would no longer come forth from the suffering of cold, nor mingle with the frosts as they fall; a country where a man can labor with the hope of someday being lord of the soil; where a man can be a man, not a dog; a country where now and for many years to come there can be a glorious investment of a small or large capital safely made; a country where, having myself resided for six long years, traveling it over and over, I have never yet seen a public beggar or a person who lived by appealing to the sympathy of others. It may be that others have but I can safely say that such people are hardly known in this fortunate country.

A country, in short, where to my knowledge there are plenty of men who are worth from ten to one hundred thousand dollars, and who, but a few years ago, had hardly a dollar they could call their own, and many of whom, had they remained in their native countries, would most undoubtedly have remained there forever poor, only getting their daily bread with their daily labor, instead of being as they now are, among the wealthiest of this naturally great and delightful country, doing what they do at their leisure or pleasure.

TEXAS

Now, kind reader, unfold, if you please, the map of the United States, and upon its southern portion you will find a country or state that, for some reason or other, is called Texas. Here is the promised land; here is the Lone Star; here is the Australia of America.

The Place to Live

It may be thought that through views of advantage, for the gratification of fanciful display or some reason or other, the writer is overdoing this thing and placing this section of country far beyond any real excellency or advantages that it possesses over other countries or neighboring states, but it is not so. Upon my word and honor, and as God is my witness, I feel from the bottom of my heart every word that I have said, and shall conscientiously continue my subject, taking with me to the end majestic truth as the guiding meteor of my pen. O how I feel, not the want of sincerity, but the want of ability, to enable me to do it justice.

Texas! Man, with all his ideas of perfection, might look in vain for a land more favored by creation and the government of man than this; and yet, comparatively speaking, her habitations are mere dots upon her surface, while in many parts of the world people are cramped up for the want of room. Here the denizen goes forth and commits to flames the vast surplus of spontaneous growth, while the people of other countries are carefully storing up every blade and spear that the soil, pushed to its greatest capacity, is capable of producing. Here the minds of men, ever so free and easy in regard to the cares of life, are revolving in schemes of speculation, and around them their cattle are cropping the luxuriant grass the twelve months through, while in other countries the eternal routine of laborious preparation in summer for the necessities of tedious winters, render there, both the minds and bodies of men forever enslaved; never dreaming that these constant cares of their lives, necessary upon a stingy and contracted plan, are robbing their very souls of their beauty; and their minds forever enslaved for the requirements of the body, never drink from the fountains of thought, which are boundless and free.

Let me introduce to our shores a stranger from across the ocean or the Gulf of Mexico. When landing, he is generally unfavorably impressed. He has cherished an idea that soon the land of Texas will stand up in ready relief to his anxious eye. He looks around in vain for that of which he had dreamed, and inquires of the country. Hearing its many praises, he loses no time in preparation for some part of the interior. He starts out over what seems to be an endless level. Soon, herds of deer are bounding away before him. He sees cattle in every direction, and wonders at the immense waste of luxuriant grass. Although not satisfied with this broad level so destitute of noisy creeks and shady groves, he begins to feel somewhat reconciled.

Now the timely setting of the sun, casting her golden rays upon the dark green foliage of the earth, reminds him of the approach of night, and that here, as elsewhere, man must rest, eat and sleep. Under the branches of a live oak or cluster of trees, not far from a lagoon or pool of water, he turns in to answer the demands of his nature. He sleeps upon the ground, in the open air, beneath the dew-catching branches, and dreams of all things glad and free. The music of the night and the first greeting of his ears at the dawning of day are the ever-changing notes of the mockingbird. He rises to welcome back the beauties of day and, being refreshed, starts on his way.

The scenes of today are those of yesterday but, toward night, in the distance, are signs of elevation. At the close of day, upon hilltops, from whence the view is delightful, that overlook the travels of the day and the sublimity of the distant plain. Nearby, the music of dancing little waters on their way below, lull our traveler to the sweet embrace of refreshing sleep.

The arms of Morpheus again relax their hold and the travel awakens to a consciousness of the rich comings that must soon follow the now gray lights of the east, happy tidings of day. The glorious orb appears and here are not only beauties for the eye, charms for the ear and sweet fragrance for the smell, but a feast for the mind. Visions of happiness and hopes of fortune are now playing their parts around his joyful heart.

Again upon his way, on a surface now hilly and undulating, the traveler is divested of the last inkling of prudence and is eager to drink deep of these moistenings of the soul. He looks and looks, wonders and wonders, and finally the tongue, nay, the very lips of the soul burst forth: "Can it be that these broad unbounded meadows are perpetually green? Can it be that the sweet fragrance of these flowers, whose beauty can only be seen with beauty's eyes, are unceasingly mingled with the air as if forever to delight the breath of man and charm the flight of the honeybee? And are these birds of so many colors forever so sweetly musical? Is it possible that these herds of cattle, horses, mules and sheep, the swiny tribe and all, live here year round without the aid of man? And could man live here throughout all the variations of heat and cold, with only the trees to shelter him, and all this in the United States of America, the renowned land of freedom? Am I dreaming or is this really so?"

Such words as these will give the reader but a faint idea of the surpassing beauty and delightful appearance of this country from early spring until summer. In the dead of winter, the stranger from a snowy country would be equally delighted; although the flowers of the forests and prairies are not so plentiful and fragrant as in summer, he is here and can realize the comfort sleeping out of doors at a time when he knows that frosty snows are covering the land of his distant home. And he can see in winter, as in summer, the cattle living upon the grass of the prairies, uncared for by man. The traveler may go on for hundreds of miles, in different directions, penetrating the mountains if he pleases, continually absorbed with the beauty, fertility and healthful appearance of the country, and be particularly carried away with the idea that the doors of wealth are here wide open to everyone who will devote his best efforts to the improvement of the advantages that lay here before him.

It is sometimes amusing to look over and consider the extravagant paper calculations of newcomers in regard to what can be done in Texas at the stockraising and farming business—i.e., as they say, and with much justness too, if men would only work one-half as hard here as they do where I came from, considering themselves proof against the habits of idleness, or *Texas fever* as it here called, to which many of the old settlers of the country have become addicted. The time was, and it is the case now in many places in western Texas, that men, by stepping to their doors or outside their camps, could shoot down a fat buck or doe, the meat of which is excellent, and the hides of which, by a simple process of tanning with which all Texans are familiar, make most durable clothes. So it is not so strange that the old settlers here, who were once so nearly cut off from the older states by so limited a communication with them, should have contracted habits of idleness, since most all they could do that then seemed useful to them was to kill deer for meat and peltry, and fight Mexicans for their independence.

For my part, I would withhold that severity of censure that the old Texan very often seems to deserve for his relaxation of energy and loose morality. I would say to the old veteran, "All honor to you for your services in liberty's cause, and happy may you live and die. But since such exertions as yours and your childrens' seem inadequate to the development of the resources of the excellent country for which you fought and conquered, you certainly will have no objections to seeing this country inhabited by those who will turn its rich soil to the greatest advantage, and its extensive productions to the good of mankind."

The state of Texas is a planting and stock-growing country. Its western portion is better adapted to the growing of stock, and not so well adapted to the growing of cotton and the great staples of the south. The valleys or bottoms of its principal rivers, however, are fine cotton and sugar-producing lands, and a large amount of these commodities are now produced and exported from the state. Their production, however, is confined principally to eastern Texas or the bottoms of the Colorado and portions of the state east of this river. Many of the river bottoms west of this, though, produce cotton, and some of them a large amount of it. But it is more the southern and western portions of the state, generally, as well as the stock-growing business of Texas, that are here to be considered.

Cattle Raising

For anything like a real picture of many parts of this country, my pen might labor in vain. I can only wish that the reader might travel it over, as I have done, stand upon its elevations and contemplate its beautiful form and the rich mantle that God has spread over it. How often have I wished that the farmers of the cold country, where I was raised, and the cold and snowy countries where I have been, could travel over western Texas, taking with them a knowledge of its climate and soil, its different kinds of grasses, and particularly a knowledge of the stock-growing business of the country. It is not so difficult for me to imagine how those lords of the soil (in summer) and of snow drifts (in winter) would feel, or what impressions of such an excursion, accompanied with vivid recollections of their frosty homes, would be. To spread out upon paper words that would justly convey my ideas of those feelings and impressions would be futile.

Would that the representatives of such feelings and situations as those that, in my beginning I attempted to portray, could only know how well this country would be adapted to their particular wants—and O that thousands of unfortunate men, women and children could exchange their present unhappy abodes for a home in this bountiful and most delightful country.

Who that has a soul, or that can feel his accountability to his God, would not open the channels of alleviation to unhappy and suffering humanity?

There is, across the deep, a country that resembles much, in many things, the one of which I am writing; the pursuits and productions of which, to a great extent, are the same—Australia. She, like Texas is a great stock-growing country. She has been more fortunate than Texas in some respects. The first cattle taken to Australia were of English breeds, and the constant importations from England have since kept up and improved the cattle and other stock of this most excellent grazing country. The cattle of Texas are inferior to those of Australia, but their vast, luxuriant range makes them thrifty and profitable to their owners. I believe they are a race of Spanish and French cattle—or I can safely say that they are generally Mexican, Louisiana and Missouri cattle. Although they are generally rough, and not compact and heavy enough, they are large and rangy, and are fine cattle upon which to improve. But, withal, beautiful droves of beeves can now be selected upon our Texas prairies.

The first sheep introduced into Australia were from those countries where improvement in livestock of all kinds has long since been a great object. Australia's adaptation to sheep and fine wool growing was finally discovered, and her importance as a great wool-producing colony soon demanded the attention of the English government, and from its patronage and encouragement it is now the greatest and finest wool-growing country in the world. But is there no country that is yet to be her compeer? Has America no Australia to furnish her looms and enable her manufacturers to compete with those of England? Has America no country where wool can be produced sufficiently cheap and in such abundance as to place it beyond the control of the speculator and counteract, in its production and manufacture, the penny labor of Europe? Certainly she has; and more than this, she has a country that will soon furnish her millions of consumers—the poor, laboring men and all— with excellent beef and other meats at reasonable rates; a country to produce her mules—that most excellent cultivating animal—so reasonably low that they will be within the reach of all her farmers; and this country, or part of the United States, is Texas, and particularly western Texas.

There are now about two millions of cattle in Texas. Their average value per head is about six dollars. The raising of beef and veal cattle is an extensive and profitable business in this country. At this moment I am thinking of a man, an Irishman, who has twelve thousand head of these cattle. He lives upon the San Antonio River, and the bulk of his cattle range between that and the Mission River, or over a range of country, probably, about thirty miles square. This man, I am credibly informed, not many years since followed the business of saddle-tree making, and has, by continued good management and economy, increased his stock of cattle from a few head to this large number. He has, of course, bought cattle, but they were bought with the proceeds of his beef cattle originating from his first little beginning of stock that he made at the saddle-tree business. He has also a large amount of other property, all of which he accumulated from this small beginning of stock and the proceeds of his saddle-tree business. He now thinks of disposing of a share of his other property and putting the proceeds thereof into cattle. Should this be done, within a short time, he will have about fifteen thousand head of cattle, and other property necessary for their management to the amount, probably, of ten thousand dollars. I say necessary, as it will eventually be required—probably three thousand dollars worth of horses, lands, etc., would enable this man to carry on his business at this time, giving all necessary attention to his cattle.

I will venture to say that if this person were asked what would have been his situation at this time had he remained in Ireland, his answer would be, a poor laboring man. He last year sold about five hundred beef cattle, principally at eighteen dollars per head. He has this year sold eight hundred head of beeves at sixteen dollars per head. He will brand this year (1859) about three thousand calves. He probably would not take less than eighteen or twenty thousand dollars for his next year's increase. Not one of this large number of cattle does he ever feed a spear of hay or anything else, nor has he any of them inclosed in pastures. His cattle are scattered over a large range of country, with the cattle of other stock growers, and outnumber those of any other brand in his range. *[Editor's Note: The above no doubt refers to Thomas O'Connor, whom T. F. Buck would have frequently encountered during his cattle purchasing and driving runs. His knowledge of O'Connor's operations implies a familiarity with the cattle baron.]*

Stockmen in this country, for the purpose of marking and branding their calves, collecting their beeves, and working with their cattle generally, go together on horseback in numbers sufficiently large to enable them to work to advantage. It is not unusual here to see as many as three or four thousand head of cattle brought together in one bunch for the purpose of selecting out such as may be wanted.

Let the reader imagine himself upon a broad prairie, covered with scattering cattle as far as the eye can reach, and from twenty to thirty men and boys, mounted upon Mexican horses, swiftly surrounding thousands of these cattle, bringing them together with the cracking of whips and whooping of voices, or Indian yells, and when together the clashing of horns, roaring and fighting of bulls, and the dextrous riding maneuvers in separating such as are wanted from the main herd, make up an interesting and exciting scene. These are everyday occurrences here in Texas, saying nothing, at present, about the lassoing of wild horses, cattle, and other animals.

Men having large stocks of cattle are not particular about bringing them home. There are generally company stock pens built over the range, and they brand at the first convenient pens, private or otherwise, anywhere within the range, then turn loose, then collect, again and again until their trip is made or their work completed. On returning home, however, from a stock hunting trip, there is generally a goodly number brought to the ranch or headquarters of the business for the purpose of branding or haunting, particularly if the drove be made upon the outskirts of the range. Should a large stock grower find his animal out of his range, or beyond certain boundaries, if he does not take it home, he works it back within these boundaries before turning it loose again.

Men of small stocks of cattle are generally more particular about keeping them near home, and sometimes those of considerable stocks have them haunted to a small scope of country. Having myself, last year, bought, for a shipper, the beeves from a stock of about fifteen hundred head, belonging to a man who was living in the valley of the Nueces, I am knowing to the fact of his having his whole stock of cattle, except, perhaps, about twenty head, brought together in one herd in less than four hours' time, and this, too, by three or four Mexicans on wages of trifling amount. Probably the branding of this man's increase of stock and the whole care of his fifteen hundred head of cattle, does not cost him more than two hundred and fifty dollars per year, and his stock will turn off one hundred three and four year old beeves, annually, at sixteen dollars per head right at home. The owner of these cattle is not able nor does he pretend to work, nor is there any necessity of his working, for his cattle will make him rich, besides supporting his family and paying for all the attention they require.

Now, Mr. Lord of the Snow-Drift of New England, Canada or anywhere else—do you think that this man would exchange his fifteen hundred head of cattle and place of residence for the best farm that could be found in any cold country, with the understanding that he should go and live upon it? No indeed, he would not. Although his fifteen hundred head of cattle are not worth ten thousand dollars, and his house a picket hut, stuck up on land belonging to someone else, he would not exchange them, with the advantages of the range they have, and his residence, as humble as it is, for property of five times its value in a cold country, where he would be compelled to house up from four to six months of the year, to avoid freezing. He was raised among the destitute of Europe; he has experienced the frost-biting and snow-beating winds of the north, and is thankful that he has found a better climate and country. His expression is, "I am sorry for the poor devils of those cold, snowy regions."

A person having a small stock of cattle, or even eight hundred or a thousand head, need not pay out a dollar for hire, if he has a mind to ride and work with a stock-driving company. In fact I know a person who has fifteen hundred head of cattle, and by riding and working in company with other stock growers, he accomplishes his entire branding without paying out a dollar for hire; and other stock growers, I know, who depend entirely upon hired help to work and manage their cattle. Others let their cattle and give a share of the increase or pay by the head for their branding. The usual share given by stock growers for branding their increase, is every fourth calf or a dollar per head for each and every one branded. Not a few, but many, men in Texas have made fine beginnings and are making snug fortunes by branding and managing stocks of cattle for the fourth calf or a dollar for branding each and every increase. A crowd of stock boys, when hunting upon the prairie, look out for each other's cattle, and all their brands and provisions are packed upon the same horse, cart or wagon, and made hot by the same fire.

When a herd of cows and calves, collected by and belonging to different individuals, are taken to a pen for the purpose of branding the young, those designed for one or two brands, according to the number of apartments in the pen and the number of hands to work, are cut out from the main herd, those for each brand by themselves and placed in these apartments, and the branding irons already hot in the fire, handy by, the throwing of ropes and jumping of fences soon begins. Although these Texas cattle are generally quite easily managed on horseback, many of them, when hemmed in a pen and run about for the purpose of roping or lassoing the calves, until they are heated and maddened from the bellowing of their young, etc., will dash furiously at the roper. He is obliged to retreat or take the fence for a moment until the fighting cow returns and mingles with the bunch, when all hands are at their ropes again catching and throwing the calves, marking their ears, trimming them, and scorching their hides with red hot letters and figures, or whatever characters may represent the brands of the different owners.

There are many men in Texas who had, but a few years ago, their brand upon but from one to five cows, who now have fine little stocks of from fifty to five hundred head of cattle, and in many cases such stocks have grown up, altogether, from such little beginnings. It is often interesting to hear some of these wealthy stockmen relate the history of their business. They will tell you that a few years back, in 1840, perhaps, "I had but two or three cows, a few pigs and chickens, an old mare or so, and if my, debts bad been paid, I wouldn't have been worth a dollar in the world. We lived in that old shanty, and many a comfortable evening have I seen by the fire in that old fireplace. We had no land, but lived and cultivated most anywhere we pleased. We were often without much of anything to eat in those days but venison or wild meat of some kind. But my trusty old rifle was sure to knock down a plenty of that. The hundreds of cattle and horses that you now see around me and scattered over the range with my brand on them, are mostly all from these few old animals and their increase."

How often have I heard them say that to get rich in Texas a man has only to buy a few animals and then lie down; but it takes a little more to get rich. He must brand up his increase in good season, and see that they do not stray off, or are not driven out of their range. A person having a start of fifty to a hundred head of cattle, will soon grow them up to a fine stock of several hundred or a thousand head, if he is careful to haunt them to a good range, and attend to them properly, which haunting can generally be done in about a year's time. After being well attached to a good range, near to one's ranch or place of residence, they require but little attention, further than the branding of the increase. It is well known that if a poor man in western Texas can get a start in the stock business, however small his beginning may be, he is sure to get well off with but little exertion, and this beginning is so easily made that a person of the least capacity or money-making qualifications, who is disposed to go into the business, is sure to make it. In the first place, he need not expend a dollar for land, if he prefers not, and, if he be economical and at all inclined to work, not a cent for lumber. The broad, luxuriant prairies are open and free to him for his cattle, and the timbers, in many places, at his disposal for the building of shanties, log houses, and the making of rails, or whatever rough improvements he might see fit to make for the present, until he should be able to purchase land of his own. I have known a person to own four or five thousand head of cattle in Texas, without owning a foot of land in the state, turning them loose in almost any range he might see fit, and building his ranch upon unoccupied lands.

Others I know, having thousands of cattle in Texas without an acre of land, give a share of their increase for branding and the attention their cattle get, or pay by the head for branding, etc. The better way, however, for a man of small beginning, is to purchase a homestead of a few acres at least, as soon as he can, as the lands are cheap and fine for cultivation. A man of any industry can improve a little farm, the proceeds of which will support his family or self while his stock of cattle are increasing to a number that will of itself support him in a few years, and eventually make him rich. His stock should, from the beginning, however small it may be, turn off a few beeves every year, that is if his first purchase was made up of an average lot of stock cattle.

An average lot of stock cattle, of different ages, in Texas, are cows and calves, yearlings and two-year olds, male and female, as they may run, including all females in the stock, but no beef cattle older than two years, and of course of the usual size, breed, and condition; which breed, etc., along a wide scope of the coast of Texas, is pretty much one and the same thing. Back from the coast, among the hills, mountains, and timbers, and in more northern Texas, the cattle are generally more compact in body, and pretty, not so wild, and of course more easily managed by an inexperienced hand. But the breed and condition, or unkindness of cattle in western Texas, should not govern the purchaser of limited means so much when buying for stockraising purposes, upon the principle that the child must creep before he can walk.

For instance, in the purchase of a stock of cattle in western Texas, the purchaser will get, for six dollars, cows that will raise calves which will sell for as much at six months old, for veal and other purposes, as will the calf of the same age for the same purposes, that may have been raised from a cow worth fifty dollars. And the new beginner might find himself under the necessity of parting with his calves and yearlings in order to get along.

If the beginner of quite limited means and new to this country should undertake the business with improved or imported cattle altogether, he might find his little means exhausted before he should have fairly commenced, and long before he could have acquired a sufficient knowledge of the country to make his fine stock business profitable, and consequently subject himself to the danger of shoals, which, if the reader will follow me up, he will hereafter plainly see.

In the purchase of cattle, the newcomer in Texas should aim to get as many beeves, of good age, in his purchase as he can, provided he gets his stock at the customary average price per head. I have known stocks of cattle to sell here for five dollars round, including a fair average of three and four year old beeves. And again, I have known stocks of cattle to sell here for seven dollars round, exclusive of all beef cattle older than two years. The first mentioned may not have been so gentle and easily managed as the last, but to the experienced stock grower this difference in the condition of the two stocks was of little consequence in comparison with the difference in price.

I am here only trying to impress the idea that to make a successful and anything like a rapid beginning with cattle in western Texas, the man of quite limited means should not be so particular about the kind of cattle to commence with. Let him buy anything of a Texas cow that will bring a calf, and get her as cheap as he can. After he is well set up in business, he can improve his cattle, if he wishes. I have known men in the northern states to buy old, superannuated ewe sheep for the purpose of getting a start in the sheep business. After getting a crop or two of lambs from these old sheep, they would begin to drop away, and would soon die out. But the purchasers got them for a trifling amount, and were thereby enabled to commence the business, which they could not have done, to any extent, by buying young and expensive sheep of the same quality of wool.

The stranger, with a plenty of money or a few thousand dollars, may come on with his fine stock in small or large quantities, according to his means. But he who brings improved stock to Texas, should understand the country beforehand, or have someone in advance of him who does, otherwise he should pay particular attention to what the writer shall hereafter say in regard to the importation of northern or European stock into Texas. I said, a little back, that the newcomer should purchase a homestead, etc., as soon as he can. But the man of small means should he very careful about putting his money into lands. He should first make sure of his stock, of whatever kind it may be; and as soon as he finds himself able to make a small outlay for lands, let him do so, and improve his place as best he can. Of course he should know, before getting his stock, where he is going to keep it; but it does not follow that he should have expended any considerable sum for lands or a place to live. *Vice versa*, the moneyed man who embarks in the stock business in western Texas should have his lands and place of business beforehand, particularly if it is his intention to raise fine stock or improve the stock that he may purchase in Texas.

One great advantage of this country to strangers, is that they may come here with a little money and put it into livestock without making an outlay for a home or a place to keep that stock; which outlay would of course take up their little capital, comparatively, in an unproductive way, where it could not afford that easy means of support, and at the same time be accumulating in a compound manner, as money does invested in stock here in Texas. A person can, in this climate, pitch strong and durable tents, or erect temporary picket or log dwellings, upon unoccupied lands, sometimes with and often without the consent of the owner. This enables him to invest his little capital as above set forth, and gives him time to look up a place for a permanent residence, and a range to suit him for his cattle or stock, also time to get ready to make a payment upon a place when and where it may suit him to buy.

I must not forget to say that it is, in one of the broad prairies that I know of, expected by stock growers there that a newcomer will purchase a quantity of land in this prairie before turning loose a stock of cattle upon it. But no one, so far as I know, regards the expectations of these people. I am knowing to the fact that men from all parts are turning loose their cattle in any range they please in western Texas, whether they have lands or not. But the better way, as I have said, is to get a little land as soon as possible, as it costs but a trifle, and can, as a general thing, be bought upon most favorable terms. If you are a man of quite limited means, put your money into stock, and then get your land on time. Should you then find yourself pushed to make your payments, sell out your male cattle, of any age, your yearling mules or mutton sheep, etc., to the best advantage you can, and keep your females and best producing stock, which will fetch you out all right.

I will here say that the best way for a working man of small means, who comes alone to the country and is unacquainted with its peculiar ways of stockraising, to commence the cattle or horse and mule raising business in Texas, is to look up some stock grower already in the business, in whom confidence can be placed, and not only consult and advise with him upon the subject, but work your stock with those who work his stock, if you can. Whoever comes to Texas to engage in anything pertaining to the stock or farming business, should move slowly and carefully at first. The difference in climate and the flattering appearance of the country, is apt to carry a stranger away with some foolish notion, and too often it takes him a long time to recover from the effects of his folly in the beginning. Strangers in this country should remember that there are already many settlers here, and that they are not all lazy and wanting in good sense, and that it would be well to find out a little how others do before "pitching in" too strong.

The cattle business, as it is now generally carried on in Texas, requires a vast deal of horseback riding, which is hard work, at first, for one who is unaccustomed to this kind of exercise. But after following it for one season, almost any man will endure a day's ride or work at the business with but little fatigue. As hard as it may be, when compared with the splitting of rails, hoeing of corn and potatoes, mowing and pitching of hay, cradling and binding of grain, etc., etc., it might justly be considered pleasure-riding. The hardest work attending the business of cattle raising in Texas is the marking and branding of the increase. But there is a kind of excitement connected with this part of the business, as there is, also, with the riding part of it, which makes it what the writer would call laborious amusement, in which every little boy and thrifty young man seem eager to engage and participate.

When stock-cattle change hands, to be kept in the country, the original brand is generally put on to them again, which is called counter-branding, and the brand of the purchaser is also put on to them. The brand of every stock grower must be recorded, to enable him to hold it in law, or in case of dispute. Men having large stocks of cattle generally have their brand recorded in several adjoining counties. They also like their cattle to go some ways from home, for the purpose of changing range and crossing with other cattle, which is conducive to health and thrift. The opposite course of hemming them into coves and bends of rivers, upon islands, etc., has a tendency to degenerate and run the stock down to puny, inferior animals. A person can keep his stock near home, however, and avoid all this by changing bulls, which costs him nothing, and by driving his stock, occasionally, away from their old stamping grounds, all of which, in time, with a little watching, are sure to return. This precaution, however, is entirely unnecessary until a stock has become very large, and has been kept upon the same range for a long time.

An immense number of cattle produced in Texas are used for oxen, and many of them are driven north. There are many thousands of beef cattle now annually driven from here to Chicago and different markets north of Texas. It is a singular fact that an inferior beef is here grown to four and five years of age and then driven thousands of miles to a northern market, and this, too, at a profitable rate for the producer as well as the drover. New Orleans is, however, the principal market for the present great surplus of beef and veal cattle produced in western Texas, but the time will come when Texas will produce more beeves than is now consumed in all the great markets of the United States. In the opinion of the writer, the time is coming when the same extent of territory in no part of the world will produce more livestock of all kinds than will the present domain of Texas.

The man of means, it is needless to say, can at this time, and for many years to come, do more with his money in Texas than he can do in almost any, if not quite any, other part of the world, particularly at the stock business. With ten thousand dollars a person can buy in Texas a thousand head of cattle, a league of as good land as there is in the world, and establish for himself a comfortable and respectable home in the way of dwelling-house, etc., all of which property, in time, would be worth several hundred per cent more than it would now cost him.

Inclosing lands for pasturage is quite extensively commenced by men of means in some parts of Texas, but the stock business here, as a general thing at this time, could not be made profitable by inclosing the stock of the country in pastures. In fact, from the nature and present state of things, such a course could not be generally pursued, and therefore our broad prairies are to be open for many years to any and all who may see fit to turn their stock upon them. Eventually almost every stock grower in the country will require a pasture, but the time is far distant when the great bulk of Texas range will be fenced up—possibly fifty and possibly a hundred years may not see this. The time is coming when the cattle and different kinds of stock in this country are to be generally improved, and the better grasses are to be cultivated. Then it is that fencing will come into general requisition. It will not cost any more to pasture a Durham bullock than a Texas animal, and the Durham, all things considered, is undoubtedly worth double the money. Who doubts, that if the present stock of Texas cattle were entirely of

English breeds, it would be worth twelve dollars per head all round, instead of six dollars?

When a great central railroad is put through Texas, from New Orleans via San Antonio into central Mexico, then look out for improvements of all kinds, and the going up of property in these parts. The time for the building of such road of course no one can now safely predict, but that such a road will someday be built, the most natural appearance of things renders certain. There are already many miles of disconnected portions of it built.

Stock Driving

The driving of stock in this country is a peculiar business. All the native stock, even sheep, are liable to *stampede.* Stampeding is the sudden starting and running of a drove of animals from being frightened; and no driving manager, however careful he may be, can always avoid a stampede. Things will very often happen to stampede a drove of Texas cattle that cannot be foreseen—such, for instance, as the unexpected raising up of a man out of tall grass in front of the drove; a first and unexpected quick clap of thunder and flash of lightning; or jumping up of a hog near to or in the midst of the drove; also many things that watchful and attentive drovers can avoid will sometimes stampede cattle and cause them to run a long way before stopping. A drove of cattle once well frightened and hard run, will afterward start without apparent cause. So it behooves the drivers to be cautious, for there is no regular rest with a drove of Texas cattle after they have once badly stampeded, until they are well quieted and broke in again. These remarks apply more to droves of beef cattle than to stock cattle, although Texas cattle of all kinds will stampede.

I was told an amusing circumstance by an extensive stock manager in this country. He says he once took a newcomer, a northern man, on a driving trip with him. Their cattle having stampeded in the night, every man jumped for his horse. The green man, who was an active, zealous fellow, mounts the first horse which he comes to, which was one that had been hoppled when turned out in the evening, when the day's work was done. After rounding in the cattle, and running their horses, now and then, to the best of their speed, for two or three hours, to hold the cattle together and get them where they wanted them, the crowd of drovers all got together, and the verdant vaquero, who had gloriously done his part, tells them that the horse he was riding had the damndest gait of any horse that he had ever ridden, and from his description of the gait, some one of the crowd suspected that Yankee's horse must have been running with his hopples on. It being dark, the suspicious man gets down, and, upon feeling, sure enough the fore legs of Yankee's horse are tied together, which accounted for the awfulest jumps, as he said, his horse had been making all night long. Of course the jug of Dexter's best, which had several times passed around camp during the evening, had nothing to do with Yankee's case.

Catching Wild Cattle

There are forests in this country where horned stock, from being neglected, have run wild. Crowds of stock boys sometimes go, with their dogs, and hunt these cattle out of the timber on the prairies, and catch or shoot them down to prevent them from tolling their gentle stock away into these forests. I know of stock boys who are in the habit of going upon moonlit nights, with plenty of help, each man with several ropes to his saddle, and as still and adroitly as possible, get between these cattle and their hiding place when they are feeding on the edge of the prairie. Then the boys rush suddenly upon them and push them in a body far out into the open prairie, and hold them all together if possible, and if not, when the captain of the crowd shouts, "Rope, boys!" every man secures his animal and ties it down. If he has time, he ropes another before they reach the timber. All being safely tied down, they are left until morning, when a gentle herd of cattle is driven to each one and they are let up and into the herd and driven away to be branded, according to a previous agreement of the crowd of adventurers.

There is another way of getting hold of these wild fellows, and that is by tailing them, which is a lively business. I was once with a crowd of stock boys when we came, unexpectedly, upon a bunch of wild cattle, some distance out from their thicket or place of hiding. The boys made after them, and one of our number caught and threw, by the tail, a fine heifer. Getting quickly down, he put a forefoot of the animal over one of its horns, marking its ears with his jack-knife, for in case it should get up and run off, it would be his. Mounting his horse, he dashed away and overtook the frightened bunch just in time to tail another before it made the thicket. This one was thrown to the ground, but for some reason or other it got away.

This tailing business is nothing more than riding up behind an animal and catching it by the tail, and sometimes giving it a twist around the horn of the saddle and, driving the spurs to the horse, he jerks the animal to the ground with such velocity that frequently it will not attempt to get up for several minutes. The faster an animal is running, the more easily it is thrown to the ground. When driving stock-cattle, if an animal is not inclined to drive well and is continually breaking out of the herd, some driver takes after it and gives it a fall by the tail, after which it generally drives peaceably along.

Camping & Storytelling

The stock drover's camp is generally replete with stories in regard to horsemanship, lassoing maneuvers, wild horse and cow catching, etc. When in camp with these gentlemen of the pitching art, I always try to keep my end up. Upon one occasion, when in company with about twenty of these good-natured fellows, after hearing several well-colored stories in regard to mustang riding, etc., I came out with a story in regard to an old Dutchman and his son, Hans, who lived not far from where I was raised, and who were in the habit of raising horses, or, in other words, kept a Dutch dairy.

Hans was in the habit of going out in the summer to work on wages, and in the winter would come home and help keep hot fires, drink cider and break colts. Hans, one day, took a colt from the barnyard and, after leading him with the bridle out into the deep snow, without a saddle, of course, mounted him and, after repeated efforts, could not ride the colt to his satisfaction. He turned the colt back into the yard.

The old man, seeing this, came out in bad humor and interrogated Hans: "What for, Hans, you don't sthick to dish golt? He musht pe proken, und might as vell pe proken today as any dimes."

"Vell, dad, he ish te Difle himself; I can do noting mit him."

"You vagapont, Hans, you mus'dt dink zider gost noting. Tish mighty fine to vorm yourself py mine fire, und to notings for your grubs. Get te pridle. You *musht* ride him, Hans."

Getting the colt, Hans succeeded in riding him, but could not get the colt to go ahead as it should. The old man, coming up on this scene says, "Hans, I has von contrivance by vich I can make dish colt go ahead. I vil get on to de gold, and you go long town pehind dat sthump, and when I rides long py dare, you trow your hat and halloo Boo!"

The old man with much difficulty reaches and is passing the stump, when here comes the hat, with a frightful jump and "Boo!" from the top of Hans' voice. The colt, of course, with a tremendous jump, keels the old man heels over head and buries him in the snow. Hans quickly approaching and shaking him says, "Dat, am you kilt?"

"O no, Hans; put dat vas too tamn pig a boo for a little golt!"

It is needless to say that, after this, I was called upon until my fund of stories was exhausted.

Horse & Mule Raising

The stockman in western Texas has a wide field in which to operate. If he has a speculative disposition, there is, and is to be, plenty of room for its gratification. If he has someone to look to his stock, they can be doing just as well as though he were all the time at home watching them. There are any number of stockmen, merchants and speculators in Texas, who have made many fine fortunes by driving horses and mules from Mexico to northern markets, or by bringing them here to sell to northern speculators, and to our stock growers or the planters of eastern Texas.

I have known boys of fifteen and sixteen years of age to go on wages with old drivers to Mexico with the privilege of investing what little money they might have in horses or mules, and put them with the drove of the employer. Many men who have started in this small way have become wealthy, large operators. Many of these drivers, after selling out enough horses, mules, etc., to pay the entire purchase money and all expenses on their droves, have fine lots of mares left on hand, which they keep here for the purpose of raising horses and mules, which, by the way, is no trifling business in this country.

It is estimated that the horse kind in Texas now numbers about a quarter of a million, but the continual driving of this kind of stock from Mexico to Texas, and from Texas to different parts of the United States, renders it very difficult to come at any correct knowledge of the number of this kind of stock really belonging in Texas. I should suppose that the tax list of the state did not cover more than three-fourths or four-fifths, at most, of the horse kind really belonging in it; and whether it does or not, the raising of mules and horses in Texas is an extensive and profitable business.

I am now thinking of a Dutchman who has made a snug fifty thousand by buying and driving horses and mules from Mexico to different states of this Union, and by raising cattle, horses, and mules in western Texas. He is now settled down in western Texas, loaning money at such rates of interest as could be paid in no country but this. I have been told that a mare colt, for which he paid five dollars in Mexico, he sold in Illinois, when grown, for one hundred twenty-five dollars; and that a mare, for which he paid in Mexico, five dollars, when on her way to Illinois, being kept over one season in Texas, brought him two colts, one soon after the purchase and the other the next spring, and all together brought him one hundred and forty dollars. Who wonders at his making fifty thousand dollars?

I am an eyewitness to greater doings here in western Texas than I have just mentioned. I know of men who have mares that were thrown in for nothing, as sucking colts, when their mothers were purchased in Mexico, and some of these mares have had their fifth and eighth colt since brought to Texas, and bid fair to have as many more. Their keeping has never cost a dime.

I have knowledge of men, and there are plenty of such here, who have made large amounts of money or property by driving horses and mules from Mexico, selling out such as they deemed advisable and keeping the balance. They are now raising horses and mules at such paying rates, upon the capital invested, as was never heard of, except in Texas. Not a hundred miles from my place of writing, there is living a gentleman who, not many years since, came to Texas with a small amount of means, and commenced the buying, selling and raising of mules and horses. I am told that he now has an increase of one thousand mules annually from his stock of mares.

Think of this, reader! A stock of mares that will raise a thousand mules annually, running on the big pasture of Texas, for nothing! The owner of this stock has recently built a home that looks to be worth fifteen or twenty thousand dollars. He could probably build one every year without any strain upon his income. Here is a country wide open for others who may wish to take his track.

This stock, like horned cattle, find here the year through more grass of spontaneous growth than it requires, and is managed with very little trouble and expense. I have known a man in western Texas having four or five hundred head of horses, mules, jacks, etc., without the help of more than one man to take care of these animals, and this help only a part of the year during branding season. This man, not many years ago, worked not two miles from the cluster of trees under which I am writing, for a single cow and calf per month, worth at that time five or six dollars. He afterward went to Mexico with the old gentleman for whom he had worked, and invested his small earnings with him. He continued driving horses and mules in this connection for some time. It was but a few years, however, before this laboring man had got more property than his old employer, he proving the most active and expert at the business. He is now settled here in western Texas, upon a tract of twelve hundred acres of fine land, which he owns, in addition to several thousand acres besides, raising mules and horses almost exclusively as a business.

This is but one or two of the many instances that I might relate to prove this a remarkably good business.

A stallion or herder placed with a number of mares, and kept with them altogether for one season, will never after suffer any of these mares to leave his herd, unless it is from some accident. Mares are kept together in this way, in different bunches, for the purpose of raising mules (jacks not answering the purpose of keeping them together.) Horses fixed and used for this purpose are called herders, and it is amusing to see them lay their ears and whip their mares into the herd when they get a little scattered out, or when a person is approaching them on horseback or otherwise, and woe to the stallion or gelding that attempts to intrude or make his company agreeable to any of this bunch of mares. The herder whips him, if he can. They do not fight jacks, but are a benefit to the jack that may be placed in a herd with them. These herders are a great help to the business, and will continue to be so as long as horses are raised upon broad unfenced prairies. The fact of a single herder keeping together fifty or sixty mares, on a broad prairie for a whole year, without ever suffering one of them to leave his herd, should be sufficient evidence of their utility, and that they will do this for a jack—submitting to Mr. Long Ears as the king of the harem—is a singular and striking proof of their utility in this open country. Horses are more inclined to stay together and ramble less than cattle.

A person having twenty or even five good brood-mares, and a proof jack, by selling his mules at a year old and putting the proceeds thereof into mares, will soon get a fine stock of breeding animals on hand. A good average lot of yearling mules, from Mexican stock raised in Texas, will bring thirty dollars round; a lot of half-breeds of this age would bring fifty dollars per head; a lot of common Texas mules, at three and four years old, from sixty to seventy-five dollars; a lot of half-breeds of these ages, from seventy-five to one hundred and twenty-five dollars, right at home; and mules of three and four years old, from American stock, are worth from one hundred and fifty to two hundred and twenty-five dollars per head.

A pretty good average of Mexican mares can now be bought at twelve to sixteen dollars here in Texas, without going to Mexico for them. There is a certain point in western Texas to which the Mexican horse trade tends. A person anywhere in this country wanting Mexican mares, jacks, mules, or horses, can go to the old town of Goliad, at almost any time, and there meet with Mexicans or speculators having such animals for sale. They of course cannot always be bought there as cheap as in Mexico, as the men who bring them there generally do it for a profit, but frequently there are great bargains made out of the Mexicans at this point.

Breaking of Texas Horses

The breaking of Mexican or our Texas-raised horses to ride, is a wild and exciting business. When first mounted they are inclined to pitch, and a majority of them do their best to get the rider to the ground.

Pitching, as I would describe it, is first putting the head down, then one end up, and then the other, advancing forward every up and down, sometimes inclining this side and that, generally with the fore legs straight and stiff, and the head always down, contracting or gathering at every jump—a furious, rapid succession of motions that soon exhausts the horse or downs the rider. An experienced or professional rider is, however, almost certain to sit firmly in his saddle or hang on until the horse gives up from exhaustion.

The saddles of this country are of peculiar make—of course well adapted to the riding of these horses. The Mexican manner of subduing these animals is adopted, to a great extent, by the people of Texas and those who come here. These Mexican or unimproved Texas horses are to be useful, and I might say almost indispensable, in this country so long as the stock of the country is left to range over unfenced prairies. There is no other kind of horses that would answer anything like as well for the stock business. Many of these Texas horses are beautiful pacing ponies, and of a variety of colors. When well broken, they make fine riding horses for ladies and children, and we in Texas think they are first rate riding horses for men.

Mustanging

Mustanging is among the wildest and most excitable games of the country. I have been told by different individuals that they have seen as many as fifteen hundred to two thousand mustangs together upon a western Texas prairie. Mustang pens, for the purpose of catching these wild animals, are built in almost all the extensive prairies of western Texas. These pens are strongly made, to the height of ten and twelve feet, and sometimes with wings from a half to a mile in length. They are built where these animals frequent and, if possible, over their extensive trail, among bushes or behind timber, where they cannot see their danger when approaching the pens until they are too far within the wings of the pen to retreat, unless it is by running directly back upon their pursuers, which they sometimes do when coming in sight of the pen. But generally, when this far, they furiously take their cage and not infrequently with such force that the ones in front, unable to stop from being pushed on by those behind, are smashed down, striking the fence, and, like a train of cars in collision, pile upon each other, until some of the hindmost ones flounce along the pile over the top of the pen. When once in the corral, and the gap safely secured with cross-poles, ropes, etc., boisterous work begins. Here, the vaquero can display his skill to his entire satisfaction.

About twenty miles from my place of writing, there is a family of young men who, several years ago, built a mustang pen which I have several times seen. These young men, having corralled several little bunches of horses and having made several ineffectual attempts to pen a certain valuable gang of mustangs running in their prairie, had abandoned the idea for a time. Soon after, though, there came news to them that a couple of little neighboring boys, on horseback, had accidentally come upon this very lot on mustangs, on the old trail, in the very mouth of the wings of the pen. Dashing upon them and bravely pursuing their advantage, these boys forced them into the pen and securely closed the gap behind them. Immediately all hands were on tiptoe and soon, with a preparation of ropes and lassoes, revolvers, clogs, hopples, blinds, etc., away they go to catch and subdue the valuable ones, which are the mares and colts, and shoot down the worthless, which are the old stallions—worthless from being too obstinate and headstrong to pay for breaking, and when broken, disagreeable animals.

I believe there were about forty mares taken from this bunch and gentled, which sold for twelve dollars per head, after they were manageable in a herd.

The gentling of these wild animals is a wicked business. It is often done by tying one forefoot of the animal to a block of wood or a weight that they cannot take away, and then left to feed upon grass, being led with a Mexican halter once a day to water, and worked in this way until sufficiently gentle to turn all together with clogs and hopples upon their feet. When manageable without the clogs, etc., they are taken off, and the herd gently and carefully worked in order to restore the flesh they have lost in the hard and tedious operation.

TEXAS

The Kind of Men for the Business

Now in regard to the kind of men who are calculated to carry on the business of raising horses and cattle in this country. —Upon opening the subject of stockraising in Texas to men of the northern states or European countries, they would most undoubtedly give one to understand that they consider themselves entirely unfit for a country like Texas, where the cows and horses are wild and are to be caught by throwing the rope or lasso, and managed of course by men who are acquainted with that art. But this would be a mistaken idea.

There are men here from all parts of Europe and America, who are engaged in the stock business, and it is hard to say who are the most expert with the lasso or rope, or who are the best managers and workers with the stock of the country. I can say, however, that I have seen plenty of men here, both from Europe and the northern states, who could throw the rope with more certainty than any Mexican I have ever seen, whose business it always is from his childhood up.

The knowledge of roping and managing the different kinds of stock in Texas is easily acquired. In proof of this it is sufficient to say that most of the men of Texas are from those states and countries where the use of the lasso is unknown, and where all kinds of stock are gentle; and these are the men who are raising cattle and all kinds of livestock here, and succeed so remarkably well. "Where there is a will there is a way." I have heard old Texans remark that, from those countries where cattle are driven by men on foot, come the best drivers and most prudent managers of Texas stock. I do not say the most expert in every respect, but I might say the most profitable. They are raised to know the necessity of working carefully and mildly with stock of all kinds. To be patient, attentive and industrious with cattle or stock of any kind, in any country, is essential, and for this reason those men who come to Texas from those countries where cattle are gently and kindly managed, will help to moderate the roughness, wildness, and carelessness of the country in the management of its stock; and furthermore, it is through these newcomers that its stock is to be generally improved and made valuable.

I would not lead anyone to suppose that they can come and manage our Texas cattle and horses without adopting the ways of the country to some extent; and these ways, as I have said, are easily acquired by the stranger, particularly the Americans and most Europeans. I am well aware that my accounts of wild doings in Texas would have a tendency to deter some men from coming here, if they were otherwise disposed to come; but I must here do away with wrong impressions.

To carry on the stock business in Texas, there is no necessity of a person engaging in the business of wild horse or cow catching unless he prefers so to do. He need not brand his own increase, even. There are always plenty of stockmen who are ready and willing to brand anyone's cattle or horses at a reasonable price, and sometimes unreasonably low. He can also, as I have said before, work his cattle with other stockmen, and, if he is determined, can become very well qualified to manage his stock in one season. What one man can do, generally speaking, another man can do.

There are many people in many of the over-populated countries of the world who would be glad to emigrate, and would do so were it not that they look upon themselves as unequal to the emergencies of the undertaking. Thousands of men never know their capacities from the simple fact that they never try themselves, and consequently live a life of comparative nothingness. Is it not right for any vigorous young or middle-aged man to believe that if way off yonder people are living and doing wonderful things, he could there live and do just such things? It is right for a person to give way off yonder matters a common sense looking into, so far as he conveniently can, at least. If the news of a newly discovered gold region reaches his ears, might he not say to himself, "What are the prospects? How extensive is this region? What proportion of the men there are making anything, and how much are they making? Is it a risky lottery of one chance in a thousand? Do a majority of the men there dig for gold or for nothing? And am I not as able to dig and go hungry, and endure as much hardship and deprivation for the sake of gold as anyone of my inches? Pa and Ma are good people, but am I always to stay at home with them and all the children? Who knows what I can do until I try?"

"So in regard to Texas. If men are there from Ireland and Germany with more property than a thousand such men would have ever made where they went from or where I am living, why could I not go there and make money as they have made and are making it, although the cattle there do require roping? If in one season this art can be learned, and I could manage my cattle by working with others in the meantime, or as stockmen there work with and help each other, why could I not do as well there as anyone? People are there from Great Britain, Germany, France, New England, and all the world, and am I not as smart as an Englishman, Dutchman, Frenchman, Yankee or anybody else? Why should I not be benefited by the spontaneous growth of that beautiful country, and my cattle live upon that waste of grass, since it would cost me nothing?"

Or another might say, "I have a fine capital, and could go there and buy cattle and pay for the attention they might require; speculate, in company with some experienced person, in mules and horses, wool and hides, and other products of the country; look up bargains in those leagues and immense grants of unoccupied lands; and live and enjoy myself in that delightful country, where a man can do more as he pleases and not live subject to so much criticism, slander, and restraint, as in this old, worn out country, where people are too nervous, too meddlesome, too artificial and unnatural for my purpose."

I will here say that a person, if he wishes, can buy in western Texas small stocks of cattle and horses, which are as gentle as they need be, or as much so as they are in any country where they are not stabled and fed with hay and grain. Many stockmen in Texas get up a large share of their calves and milk the cows for a little while, for the purpose of gentling the calves and keeping the cows near home, which has a tendency to keep the stock from scattering. I have known men to milk several hundred cows during the season a short time each, and it is not an unusual thing to see five hundred or a thousand head of cattle with bunches of mares and mules among them in front of the stock grower's pen or place of branding, congregated there from off the prairie for the night. And the reader can judge something of the wildness or gentleness of these animals when I tell him that he can walk among them all, and put his hands on many of them, and some of them, when lying down, require kicking to make them get up out of his way.

The reader must not think, from my remarks, that it is all wildness here. It is as much the way of managing the stock of the country, whether wild or gentle, that would be new and curious to the stranger as anything else. The ways of the country once learned, the cattle and horse business is many times as profitable, and a thousand per cent easier, than in any snowy, cold country in the world.

Hog Raising

My remarks upon the stock business should have continued directly on from horned cattle and horses to other kinds of stock, without going off upon seeming irrelevancy, but from the fact that other kinds of stock can be managed here without the lasso or rope, and that a person from any part of the world can manage it to advantage from the beginning if he knows how to manage it anywhere else.

Raising hogs and making pork in Texas is a business that many men would prefer to any other stockraising branch. There are vast portions of this country that are covered with different kinds of oak, pecan and other producing timbers, and many are the thousands of hogs that come to the call of the horn out of these timbers and the prairies around for their bate of corn.

It is contended by many that there is nothing so profitable as the raising of hogs in Texas. It can be done upon a small or large scale with equal success. In this business it is necessary to raise a field of corn for the purpose of feeding your hogs sufficiently to keep them haunted wherever you may want them, and for the purpose of feeding the sows when their pigs are young, although some people pretend to raise hogs in this country without ever feeding a kernel of corn or anything else. But frequently calling them by the sound of a horn or the voice to particular places, and feeding them sufficiently to keep them gentle and from running away, is by far the most profitable. Unless the love of ease in the shade has overcome their inclination to work, men will always raise corn for tolling their hogs, at least as the most agreeable way of managing them.

As I am not in a hog-raising section of country, and as my personal acquaintance with the business here is not very extensive, I will close the subject with very few remarks. I can say, however, that I have been well acquainted with the business in other countries, and have seen enough of it here to enable me to say much more in its praise than I shall take time to do. To relate the best I have heard as having been done in Texas in the way of raising hogs and making bacon, I fear might injure my cause, for I know it could hardly be credited by those who are unacquainted with the country.

The constant emigration to the country, and the labor and attention that this business requires more than other stock pursuits, and the fact that other kinds of meat are so easily raised or obtained, occasions the neglect of this branch and renders pork in great demand at extremely high prices. Although there might be the greatest abundance produced in this country, at fine paying rates to the producer, there is a vast amount of bacon, pork and ham imported into the state. Owing to its convenience, palatableness, etc., this kind of meat is generally preferred by the mass of people, laboring men particularly, to any other kind of cured or salt meat. All of this insures a good home market for the produce of those who may see fit to engage in this business and make such meat as is fit to eat. And I believe that there was never better pork or bacon than is made upon Texas mast. It is no more like northern or western beech-mast meat than a hard headstone is like a soft brick. The mast-fat pork of this country, when boiled, does not half shrink away, nor when fried does three-fourths of it go to oil. It is more solid and firm, like northern corn or mush-fed meat, and is very sweet.

New and extensive markets will continually open for the producer of hog meat in Texas. When the great (*is to be*) railroad shall have penetrated into central and western Texas from New Orleans, then that great city of the south may look this way for her supply of fresh pork and mutton, as she now does for her beef and veal. Working men can go into good hog-raising sections of western Texas and make a glorious beginning with but little capital.

How often have I thought, when traveling over different portions of this country, that many farmers of the northern states and the cold Canadas and many other countries, would be overjoyed to know what they could do here at this business. If men of small capital or limited means would interest the right kind of laboring men with them and come to western Texas to engage in this branch of the stock business, they could make everything around them smile.

In the first place, the land that would be required to make a respectable beginning at the business would cost but a trifle, and could be bought upon easy terms. A stock of hogs could soon be raised from a few breeding sows, which would cost but little. I know of fine stocks of hogs, numbering several hundred, that were raised in a short time, originating entirely from a few head of pigs, that cost nothing but the trouble of taking them wild from the timbers and taming them. Dwellings and improvements that would answer the purpose for a beginning, could be made of timber found nearby, with little labor and without the outlay of money. Soon the settler could blow his horn, and hundreds of the swiny tribe would come at his call. In seasons of plentiful mast, fine droves of fat hogs could be turned off to advantage, or perhaps salted, smoked and packed away to better advantage. I know of a gentleman who sold out of his smokehouse, a year or two ago, fifteen hundred dollars worth of bacon, all of which was grown and fattened upon the mast near his premises. Others I know of here carry on in the business much more extensively than this gentleman—I should say five times, perhaps, and not be far off. Should a person be determined to go into the business upon an extensive scale, and raise corn for the fattening of his hogs, he can do so with the most complete success. Early planting generally insures a good crop of corn in Texas.

In the selection of a location for the hog business exclusively, a person should of course be somewhat more particular than when looking for a range for cattle or horses. This country seems, by nature, to vary in sections, each being adapted to a particular kind of stock, although any or all kinds of stock will do exceedingly well in many sections of western Texas. The reason why men of capital should interest or bring laboring men with them to carry on the hog business is that labor of the kind required is hard to get here. This is unlike other stockraising in this respect, as it is in almost all others. Mexicans and white men, who are ready and willing, can be hired to work with horses and cattle. Mexicans can be had at a low rate for the purpose of herding and working with sheep, another kind of stock that the raising of here is considered by many the most profitable and agreeable of anything that can be done in the way of stock business.

TEXAS

Sheep Raising & Wool Growing

I might as well say at the beginning of this subject that, although Mexicans are good herders of sheep and work at low rates, northern men, Scotchmen, and Europeans generally, who understand the management of sheep, can do well here as managers, herders, and laborers with this kind of stock.

The rapidity with which the sheep business is going ahead in this country is sure to make a great demand for labor of the kind required at the business, and northern men and Europeans as herders and laborers with sheep would be preferred to the Mexicans, whose ways are too wild to suit those who are to be the great wool growing community of Texas. Another advantage that the northern man would have over the Mexican is that his language is the same as that of the generality of wool growers who are to occupy the country. Germans are to be wool growers to a great extent in Texas, which will give an extraordinary demand for laborers of the German tongue at this business. Americans and all countrymen are glad to employ Germans as herders of sheep or anything else.

The growing of wool is to many a most interesting business, and one with which the writer is well acquainted. I have a knowledge of the wool growing business of the northern, eastern and western states of the Union, and of different parts of the world, and my opinion is that nowhere can this business be made more profitable and pleasant than in western Texas. Years ago there seemed to have been a want of acquaintance with the business as adapted to the country, which, by degrees, has been and is now being rapidly acquired. There are already many instances of success in Texas that can hardly be equalled in any other country. I will here introduce extracts from a publication in the *Texas Almanac*, written by a gentleman who has had a good deal of experience in the sheep business in this country. He says:

> *In preparing a second article upon sheep raising, in this portion of Texas, I find that I have but little to add to the experiences I gave you last year; my success has continued most flattering since September, 1857. The winter of '57-8, although very wet, was passed without any loss worth mentioning; two ewes only died, and both from extreme old age rather than from any disease. At any time during the months of December, January, February and March, nine out of ten of my wethers, although running without shelter and with no other food than what they could pick or crop upon the hillsides and in the prairie valleys, were in better condition for the butcher than stall-fed animals ordinarily are at the north, and since spring set in the greater number have been too fat for the shambles. "Try and find one poor enough to kill," has been the common request for three months past, whenever I have wanted mutton for my table.*

As evidence of my success for the past two years, or since May, 1856, I will give a short statement of the increase in the number of my sheep, and in the amount of wool sheared. I doubt whether a greater degree of good fortune ever attended the efforts of anyone engaged in the business. In May, 1856, I had some eighteen hundred and fifty sheep and lambs, all told. Had I not sold or killed any bucks or wethers, I should have been able to count over four thousand at the end of May of the present year (1858). From this it will be seen that I have more than doubled my number of sheep in two years. Meanwhile, the increase in the amount of wool has more than trebled, as the following will show:

In 1856, I sheared.........2,800 lbs.
In 1857, I sheared.........5,100 lbs.
In 1858, I sheared.........9,000 lbs.

And this, after selling and killing nearly four hundred wethers, and without purchasing a single animal. When it is taken into consideration that the quality of my wool has materially improved by breeding from no other than pure Merino bucks from the best flocks of France and Vermont, it may safely be set down that, while the quantity of wool has more than trebled in two years, its value has fully quadrupled. Am I not right in saying that so great a degree of success has never attended the efforts of anyone engaged in the same business?

I cannot reasonably hope for a continuance of such unparalleled good luck and fortune; yet I can see no reason why so great a degree of mortality should visit my flocks in the future as ordinarily prevails among sheep in Ohio, Pennsylvania or Vermont at all times.

Here in the mountains of Comal and Blanco counties, I believe it to be impossible for two great scourges of flocks, almost the world over, to be generated and spread. I have reference to the foot rot and the scab. Nor do I believe that the worst of all epidemics among sheep—the liver rot—can ever cause much loss to our flocks in this high and dry region. Not a case have I seen in two years, nor can I point to any causes in the mountains to give it a foothold. We might as well look for a visitation of the yellow fever in a region where even the lightest bilious attacks are almost unknown, and where physicians are compelled to resort to other callings than their regular profession to gain a livelihood; and if we are to go on and escape the diseases I have enumerated, we have undoubtedly the best region for sheep in the wide world.

In proof, I would state that good grazing lands can still be purchased at from one to two dollars per acre, and that the cost of watching, salting and caring for a flock of a thousand head, does not exceed two hundred and twenty-five dollars, or twenty-five cents per sheep per annum. How can the northern and western wool growers compete with us on lands which they value at from twenty to sixty dollars per acre, and where it costs eight or ten times as much to feed a single animal a year? As well might they attempt to raise sugar and cotton with the hope of gaining the profits made in Louisiana and Mississippi, as to raise wool as cheaply as we can produce it in Texas.

Attracted by the heavy profits made in this region during the last two years (for it may safely be set down that those engaged in the sheep business have cleared from sixty to eighty per cent per annum on their investments), quite a number of gentlemen have started off this spring and summer in search of flocks, and others will doubtless soon leave. Some have gone to Arkansas and Missouri, others to Mexico. I am confident that those regions will be completely swept of all the surplus sheep they have to spare, and at prices at least twenty-five per cent higher than ever paid before.

[Editor's Note: The above selection was taken from the 1859 *Texas Almanac.* This was the second year that George W. Kendall, the founding father of the sheep trade in Texas, penned an article for that publication. Lengthy articles on the booming sheep trade were staples in the early *Almanacs.* Mr. Buck did not mention Kendall by name in the original 1860 edition of this text. As was considered good form in the 19th century, he simply refers to George Kendall as "Mr. K." Kendall's name is inserted in this edition for clarity's sake.]

Thus writes an intelligent wool grower of western Texas, a Vermonter, I believe, by birth. Although this gentleman does not publicly ask people of any section or part of the world to come to this country and engage in the sheep business, he does, if I am rightly informed, privately advise and even persuade his particular friends, wherever they may be, to come here and embark in it as the very best thing that they can do, or as the most safe, profitable and interesting pursuit in the world. It is an old and favorite idea with him that he would rather witness his innocent, playful lambs dance, caper and run about, than to witness the most beautiful theatrical or dramatic display. He has a taste for this business—he loves it—and he makes money at it. He has seen much of the world, but admires the country in which he lives, and not only appreciates it as it is, but sees it coming to greatness.

And why should he not advise his friends to participate with him in the profits of a most interesting pursuit? Why should not his heart swell with emulation when he thinks of his cold, frosty native region, in comparison with the mild and most inviting one of his adoption? It is but a short time since I met a young gentleman in Texas, on his way up the country with a flock of sheep. He told me that Mr. Kendall was the cause of his coming to Texas with those sheep; that he, Mr. Kendall, was a friend, and that he knew his friend would not advise or persuade him into anything out of the way. It would be a fortunate thing for thousands of cold countrymen if they were the friends and acquaintances of this gentlemen, Mr. Kendall, whose remarks I have taken from the *Texas Almanac*.

But in the absence of such a friend or advisor, may not the writer's remarks be of some service? He has himself had charge of six or seven hundred sheep in this country, though under most unfavorable circumstances. He was raised at the stock and farming business, far north of this, and has an extensive practical knowledge of not only sheep and wool-growing in the north, but all kinds of livestock. Having been nearly seven years in constant contact with Texas stock growers, managing sheep, buying, driving and shipping beef and veal cattle, branding and working with stock cattle and horses, and to a certain extent with hogs, he ought to be well acquainted with and able to give a pretty good account of her stockraising and farming business and capacities.

To continue the subject of sheep and wool growing in this fine country, I will introduce the operations of an acquaintance, living not far from my place of writing. Some years ago, I spent a week with him at his sheep range. About the year 1840, he commenced with twelve head of Mexican ewes and a buck of common stock. In the year 1854, the increase of those few animals numbered thirteen hundred head of beautiful sheep. That year, the storm that blew down Matagorda, a town not far from where these sheep were raised, and proved so destructive to our coast, by the merest accident killed about seven hundred of this flock. This accident was their getting on an elevation that was afterward surrounded by water and, when swimming off, the larger half of them stuck in a quantity of floodwood and drowned. But the increase of the four or five hundred that survived the storm have since brought the number of the flock up to somewhere near sixteen hundred head. This flock has gradually improved and now yields a good quality and fair quantity of wool, and produces the finest of mutton.

The owner of these sheep is an old sailor by profession, who was born and raised in the state of Maine. I was going to say that he would not give his flock of two thousand sheep, together with his five thousand head of horned cattle, lands and place of residence, for his entire native state, and be compelled to go back there and live out the balance of his days, but I fear this would be too extravagant an assertion to make without first consulting the old sailor on the subject.

Had not this flock met with the accident of 1854, it would now probably number four thousand and more.

This flock of sheep was not raised upon such lands as are generally esteemed good for the business, nor in what is considered a fair sheep-raising section of Texas, and the old sailor has told me that for weeks at a time his sheep were left entirely to themselves, and always without a herder. They were raised, however, upon a peculiar spot, otherwise they must have been herded. This spot, the reader, if he pleases, can take almost any map of the United States and directly see. It is on the Gulf of Mexico, a narrow neck of land or peninsula, extending from opposite the town of Matagorda down to Pass Cavallo, the entrance to Matagorda Bay.

The grass of this range was eaten down years ago by cattle, which made it better for sheep. It being narrow and nearly surrounded by water (consequently having no wolves upon it), the sheep were suffered to go weeks together unlooked for, and the only herding they got was at occasional times, when someone would go up the peninsula and bring them down for the purpose of working with them or to keep them from getting too far away from home. This neck of land is now fenced across, I believe.

[Editor's Note: The stockman in Matagorda County whom Buck describes spending time with, is Thomas Decrow, a very successful sheep raiser and member of the Old Three Hundred Decrow clan.]

It is my impression that there is no livestock property in these United States or anywhere else, all things considered, that is paying a better interest than is this flock of sheep, numbering, as I said, nearly sixteen hundred head, and fenced into a pasture where they require no feeding, herding or sheltering from one year's end to another. I don't know but the thirteen head of sheep that commenced this flock would have reached its present number, or the number it would have been without the 1854 accident, in some other countries with due attention, but to have been turned out to shift for themselves, winter and summer, as were these few head, it is doubtful whether they would have done as well anywhere else on the face of the earth. The old sailor told me that his sheep would have done much better had he been more particular about having his lambs come at the right season of the year. He says he has known many of his lambs to die from being dropped in the chilly, rainy weather of winter. The best protection that his sheep ever have are the banks of the beach, which, in some places are high enough to break the wind to some extent.

Now in conclusion about this flock of sheep. If such work as this can be done in western Texas, upon lands which, for the sheep business, nobody pretends are to be compared with the undulating, hilly and mountainous portions of its interior, is there not good reason to believe that this country is to be second to none in the world for the raising of mutton and growing of wool? And does it not deserve the attention of those who are looking for new countries most suitable for this business, with the intention of engaging in it? It seems to me that it does.

I might go on and enumerate many instances of good success at the sheep business in western Texas, but since everybody is finding out its adaptation to the growing of wool, through different papers and different channels, I will not occupy time and space in this way.

A large portion and different sections of western Texas seem to be better adapted to raising sheep than to anything else. Much of this country is high, rolling and dry, and to all appearances could not be better for the purpose in question. Experience is every day proving this part of the world to be everything that it seems to be for the sheep and wool growing business; and there are thousands of men in the southern, northern and western states, the Canadas, and eastern world, who would do well to come to western Texas and engage in this business.

There has been a good deal said of late in regard to the depression of the wool markets throughout the world. This, no doubt, should be a caution to those who are inclined to embark in the wool growing business in some countries, but not to those who would commence it in western Texas. The truth is, with the right kind of sheep, wool can be produced in these parts for twelve cents per pound, and good muttons for a half-dollar per head, and money in the business at that. But does anyone suppose that good French merino wool is coming down anywhere near this figure, or even as low as twenty-five cents per pound, to remain any length of time? And who doubts that good lambs and wethers will bring from one to three dollars per head, for years to come, in Texas? At this time, wethers and good muttons are sold for ready cash at three to five dollars per head.

There is this to be considered by those who would make wool growing their business: the sheep is a double producer in a compound sense of the word.

First—in many flocks of this country, a large number of the ewes have twin lambs, and many of them have two crops of lambs a year, (but experience has proven that it would be better for them to have but three crops of lambs in two years, or, in other words, it would be better to have the ewe recover somewhat from the effects of suckling before dropping another lamb.)

Second—the flesh of the sheep being good to eat, if there should not be a demand for its fleece, there might be for its carcass or the surplus of carcasses that the wool grower might have on hand for sale. This being the case, by marketing his mutton he might be able to hold onto his wool for a better price, should he think proper, or sell his wool and hold on to his mutton. This, in a manner, gives the wool-grower two chances to one against the grower of other kinds of stock.

But it is not every man's fortune to have a taste for the sheep business, or the patience to attend to sheep properly or in a way to make them very profitable. The hardier kinds of stock are much better for some men to raise, although there is not so much roughness and hardship attending the raising of sheep, and as an evidence of which I will here relate a circumstance.

Having been in Australia, and when traveling in the interior of that beautiful island upon which there are many millions of sheep, I frequently met with large flocks and would often make inquiries of the shepherds pertaining to the business of that country. I, one day when on a prospecting excursion for gold with a company of young men, accidentally came upon a flock of about three thousand sheep, with only a little girl and two shepherd dogs to look after them. After inquiring of the shy little girl, of twelve or thirteen years, who seemed to be almost afraid to talk to a stranger, where I could get some water, I went where she directed me, and there found a little hut and the mother of the little shepherdess, a Scotch woman of middle-age, whom the cares of life seemed to have but little worn.

This woman told me that she had the entire charge of this flock for weeks together, with only the assistance of her little daughter and her true shepherd dogs. In the season of shearing, trimming of lambs, etc., they were, to a certain extent, relieved of their charge, but, generally speaking, the whole care of the flock devolved upon the mother and daughter the rest of the year. During the day, either one or the other was always with the sheep, unless, at times when they were brought within sight of home and left with the dogs. At night, the flock was inclosed in a yard adjoining the little house of the shepherdess, in which there were many cracks and openings, but through which blistering snow never came to interfere with its comforts. I was curious to know why the care of this large flock of sheep was left to a woman, the like of which was unusual in the country, and asked many questions.

I learned that it was the discovery of gold upon the island to which she owed her situation, this discovery having created a rush to the mines of nearly all the men in the country. The flock she was watching had been left in a small inclosure by its former shepherd where it was likely to starve. The owners, who were merchants in Melbourne had been unable, just then, to get anyone else, so she volunteered her services for a round price.

"But," she said, the tears trickling down her cheeks, "I had a little boy to help me then, or I could never have thought such of such a thing. More than a year since it pleased God to take him from us, and I had no one to help me then but my little daughter, whom you saw with the sheep. Surely when my little Johnny left us, I thought I should die with grief; but being cheered up by those who were sent, for a while, to my relief and assistance, I continued on as best I could, until another rush to the mines left me again with only my little daughter. We now manage the flock most of the time by ourselves, and with the help of these faithful dogs, get along very well."

Thanking her for her kindness, and wishing her all possible good luck, I bade her good day and went on my way, thinking of the trials and strange fortunes of poor mankind, and of the courage and fortitude of this woman. What other kind of stock could she have got along with as she did with these sheep? Of course not any, which is pretty good evidence that wool growing in these mild climates is not attended with the roughness and hardiness that is required at the raising of cattle and horses, or even sheep, in colder countries.

Now if a woman, with her little girl and shepherd dogs, could do so much with a flock of three thousand sheep in Australia, what could not a man, having one or more little boys, do at this business in western Texas? Or a couple of young men, for instance, who were determined to make a fortune at the business?

It is the writer's intention to buy a flock of sheep, before long, and commence the growing of wool in western Texas. I shall go to some hilly or mountain range, and erect a comfortable shanty or two for the accommodation of my family, and then, with the assistance of my little boy, now in his ninth year, I will manage my flock, which I hope may number five hundred to begin with, but if it is not more than fifty, or even as few as twenty-five, I am determined to make a beginning, and the growing of wool in western Texas shall be my business in the future, unless Providence otherwise decrees.

Should I start off tomorrow, as I am situated, I would not think of buying a foot of land, unless I might meet with a change of buying a small tract, which should suit me for a home and a sheep range, and which I should buy (if at all) mostly on time, and upon most favorable terms. If I had plenty of money, my first lookout should be land, a large tract of which I would secure as soon as I could find one that should suit me. But as my means are very limited, I should take up some suitable place for the raising of sheep, where I might feel myself justifiable in stopping for awhile, and where, with the permission of the owner of the section, league or grant upon which I might be, I would bring my flock and carry on my sheep business until I should be able to negotiate for lands of my own. Would no such a course as this be better for the beginner of small means, than for him to get a tract of land in a stockraising country without having any stock to bring on to it or money left to purchase a beginning of some kind of stock?

The time is coming when the beginner at the stock business in western Texas will have to purchase his lands at the start. Then, of course, his money will not go as far as it now will. It will then be harder for the man of quite limited means to get into the wool growing business in this delightful country. But it will be a long time before this fine sheep region is so occupied that settlers will be obliged to buy much land in order to carry on the business—perhaps fifty years.

Judging by the past, I should say that those who engage in the wool growing business in western Texas, and have a taste for it, after learning a few important facts in regard to the management of sheep, will, in nine cases out of ten, succeed most admirably—such facts, for instance, as these: that sheep should not be let or driven out early in the morning, before the dew is off, particularly into tall grass—short grass is best for sheep; they should not be suffered to run upon fresh prairie burns in the winter season, etc., etc., all of which facts can be readily learned from men of experience already in the country at the business.

It has been the opinion of some, that in consequence of the great demand for sheep by those who have and are now getting into the business here, there must soon be a scarcity of sheep for sale everywhere in these parts. But thus far this opinion has proved incorrect. There has been for sale plenty of sheep at the old town of Goliad, all along this season, and it is believed by some that there will continue to be plenty of Mexican sheep offered at this place for years to come. This is a point where Mexican stock of different kinds is brought by Mexicans and others, and as long as there is a demand for sheep, speculators will continue to bring them on. In fact, I may safely say to anyone who wishes to come to western Texas and engage in the business, he may come with certainty of getting sheep somewhere here, should he prefer not to bring them with him. There are those here who have and others who will commence the sheep business and soon abandon it, for the want of experience and a taste for the pursuit. There will be some flocks offered by such men. I know of men here who have the wool haggled off their sheep and thrown in the dirt together like pulled wool. They can hardly distinguish coarse from the fine quality. Most of such men will give up the business to those who are better qualified to carry it on. A great many who have recently come from Vermont, New York, Ohio and other states have brought small flocks of good sheep with them by railroad and water. This makes a pretty costly beginning, but if can be afforded, it is no doubt the best thing a person, coming on to engage in the business, can do if he has been acquainted with sheep raising.

The importation of sheep and hogs into western Texas is safer than the importation of horses and cattle. It is the impression of the writer and others that the demand for sheep here will cause speculators to bring them in across the country from Illinois, Indiana, Ohio, Kentucky and Tennessee. They are already being brought from Missouri and Arkansas; and that they will be brought from both old and New Mexico and offered for sale in different parts of western Texas is a certainty. Let me assure those who would like to know it, that there are plenty of wide awake fellows who know that this is bound to be a wool growing country, and that sheep are in great demand here, and not a few of them will try to make a little money by supplying this demand from wherever the sheep can be got.

I should say here to those who are coming, come on with or without your sheep; you cannot well go amiss. It might be well for many of you to bring a few bucks to improve the flocks you may buy. I will say to Illinois, Ohio and the western states: You might as well surrender the sheep and wool growing business, and bring your stock down here. We will give a good price for it. As for old New England, Vermont and those countries where the grass grows out of the rocks, trudge along in your own way at anything you please—only, send a few more of your sheep and some of those fellows who know so well how to manage them. This, remember, is the Australia of America, where we don't have to cut and cure a hundred tons of hay to winter a thousand sheep—no, not a pound. This is the place to live—where now the Indian is no longer dangerous to the white man; where now an unsteady government and the roughness of society is no longer an objection to the country; where things have changed and have become not only endurable but pleasant and attractive; where, from various causes, the importance of the country has but recently been fully understood; and where now thousands of respectable and intelligent people, from all parts, are coming to make their fortunes and settle down for life.

Importation & Improvement of Stock

The furnishing of improved and good-blooded animals to the stock growers of Texas, with which to improve their stock of different kinds, is to be an extensive business, in which many can and will eventually operate; and I believe the men who are best calculated for this work are generally living in the north. I do not allude to any particular section. The boys of Kentucky could, no doubt, do a glorious share in the improvement of the stock of Texas, particularly her cattle and mules, and they are already here at it; but as yet they have barely made an impression. The great work is hardly begun.

I am not the fool to suppose that all I have to say must be listened to or followed as infallibly correct: I am aiming at the truth, and what better can anyone do?

I have a favorite idea, that I will here suggest, in regard to raising fine bulls in Texas. For several years I have been engaged in the purchase and driving of beef and veal cattle for a shipper in western Texas. I long since discovered that there are now and then improved cows scattered over our prairies, among the common stock of the country, and sometimes there are stocks of cattle that have quite a number of such animals among them. Their owners, as a general thing, do not esteem them very highly, as they know not their blood and value, and often do not distinguish them from their other cattle. Some of these cows are half-breeds, from Durham sires and Texas dams, some quarter-breeds, and others more or less mixed with improved blood.

I have often thought that if one had a suitable pasture, which might be of the cheapest lands in Texas, and was able, he could purchase these cows when and where he might meet them, and bring them to his pasture, and then place with them a thoroughbred animal, from which he could raise a very good quality of bulls that would answer remarkably well to turn loose upon our prairies, and commence the general improvement of the common cattle of the country. The increase of females he could continue to improve, and finally bring his stock to a state of perfection. Meanwhile, he could be offering bulls of a better and better grade, until he would finally be able to offer Number One improved or thoroughbred animals that would answer to turn upon our prairies without fear of loss from exposure.

As his cows, lands and everything but his thoroughbred animal would cost him but a moderate amount, it is plain to see he would be enabled to sell young bulls for much less than it would cost to import them, and still afford a profit. And although these animals would not, at first, be as good as full-bloods, they would be what the stock grower would not be afraid to buy, and much better for him to commence with than imported animals that might die from neglect, or which he might not feel inclined to attend to in a way to make it profitable. The partially improved cows of which I speak as being scattered over the country, are generally from animals that have now and then been brought here and turned loose, and have died from neglect or some other cause. Sometimes they are issues from someone's animal who is attempting to improve his stock. By selecting the choicest of unimproved Texas cows, even, and using them in the way above described, a person could make money at the business. The stock grower is better off to have half-breeds that are raised in the country with which to commence the improvement of his stock, than to have many of the humbugs that are imported.

Not far from my place of writing there is a pasture of several leagues, recently inclosed, now containing three or four thousand head of animals. The owner of this property, several years since, purchased a Durham bull, which he has kept most of the time in a pasture adjoining his dooryard. During the season, the bull has run with the gentleman's milch cows, which are of the commonest Texas stock. This gentleman has several crops of half-breeds from this animal and is now raising three-quarter-breeds, and next year, I understand, will have enough animals that are more or less improved, to serve all the stock in his large inclosure. If this gentleman can improve from the commonest of cows of the country, how much better could a person do with the choicest of Texas cows, or the partially improved ones to which I have alluded? The gentleman who has this large pasture is an exception in the business. I know of no other person in this part of the country who has made so extensive a pasture.

I have mentioned this case only to show that by bringing on a fine bull, jack, stallion, etc., a person can raise fine stock here in the country for sale, all things considered, to as good and possibly better advantage than he can import such stock for sale. I may be mistaken in this, for I am well aware that there are several gentlemen in the country who are making fortunes by bringing on improved stock from the north and elsewhere, which they dispose of at good prices.

I said that the gentleman having the large pasture purchased a fine bull several years since. He has since purchased several fine bulls of the enterprising Kentucky boys who are here, and is now in advance of anyone in this part of the country in the way of improving his cattle. I presume that not one in a hundred of the stock growers of western Texas have made a beginning at the improvement of their cattle as yet.

There are millions of useful animals in Texas that are not half as good as they should be, and the man who will come here with good improved stock and settle in the right kind of place, can raise from it and sell to good advantage all he can part with. I am acquainted with a young man, the brother of a noted and wealthy wool grower in Vermont, who came to western Texas several years ago, bringing with him a fine flock of Spanish merino sheep, all the way from Vermont, and selling them out directly, returned to that state and brought on not only a fine flock of the same kind of sheep, both ewes and bucks, but male and female calves, and a fine breed of pigs, from all of which he has since been raising and selling at most profitable rates. This gentleman finds it impossible to supply the demand for his improved animals from the increase of his stock, and consequently continues to bring them on from the north for sale and to keep up the excellency of his stock. He has exchanged more or less of his fine stock for mares, and is now raising mules in connection with his importing and fine stock business.

Last winter he brought on from Vermont and other states his fourth flock of sheep, this making him an extensive importer of fine stock into the state of Texas from the northern states every year, but one, since his first arrival here with a flock of sheep. His last flock was partially made up of Ohio sheep. He also brought on, from somewhere in the north, a lot of fine stock for another gentleman.

And right here, to show the reader how the importer of improved stock into Texas may subject himself to the danger of shoals—of which I have spoken—I will tell him something of the management of the gentleman for whom this stock was brought on.

The stock consisted of some three hundred heavy wooled sheep, ewes principally, and some ten or twelve young Durham bulls and heifers—they were calves, yearlings and two years old. After the first mentioned gentleman had landed this stock at Powder Horn, in Texas, from off the ship across the Gulf, it was given over to a couple of young men who were sent to take charge of it and drive it up the country. These young men, or boys, knew nothing about northern stock, and could not have known much about stock of any kind, for they started out over the prairie and so managed as to be caught out in the night, miles away from any dwelling, on the open prairie. Instead of watching their sheep, they laid down and went to sleep. The flock wandered away from their camp or wagon, and the wolves got after it and killed some twenty head of the sheep, which cost, delivered here, the snug little sum of nearly twenty dollars per head.

They gathered up the balance of their scattered flock the next day, and went on, now driving too fast, and then too slow, or sometimes laying up when they had no business to be, starving their stock around some house where the grass was all eaten out (it being the dead of winter). So managing, by the time they had driven four or five days, or before they reached their destination, nearly all of the fine Durham stock died from the effects of the voyage and the abuse of these young men who were sent to drive them to their new home.

Had there been someone sent for this stock who had considered that they voyage had injured it, more or less, and that it should be driven and managed carefully, it no doubt would have reached its destination and have done well. He paid double its value in the north, and the young men, or boys, should never have been sent for it, which proves a want of experience and attention as much as it does danger in the business. The gentleman who brought them on and delivered them to these youngsters went on foot with his own flock, and arrived safely at home in the interior without trouble or loss. He told me, when I met with him on the road with this very flock, that he could not wish for better success than he was having, or a better business than he was doing. I now hear that this fall he will go on to Vermont or Ohio and bring to Texas his fifth flock of improved sheep and other stock.

This gentleman's ranch is situated on a beautiful clear stream of water, where almost all the country in his vicinity is covered with scattering trees, the shade of which is just the thing for his imported and improved stock. The grass of his lands and range is the better quality of sedge, which is short and fine, and grows around the trees and upon the little open prairies which are skirted with the scattering trees of these timbered prairies. The timber upon his land, in places, is sufficiently thick to enable him to take out enough to fence it up into fields and pastures, and leave enough standing for the purpose of shade, firewood and other conveniences. This gentleman told me long ago that he had sold a large number of animals, getting for his bucks from twenty-five to a hundred dollars.

Numbers of such men as this could come here and settle over the country, at different points, bringing with them their fine stock, the increase of which, together with prudent importation, would make them rich. They could sell to the stock growers around them, often taking horses, mares, cattle, half-breeds or Mexican sheep in exchange for their animals, and become Texas stock growers as well as improvers and importers. Men coming here for this purpose should not be particular about bringing the most expensive stock at first, unless they might be owners of such. But they should bring well-bred if not thoroughbred stock. Bring no humbugs, as this would result badly for those who might bring and palm them off.

There came to western Texas, a year or two since, some men from Missouri with a large drove of jacks. The jacks were all haltered, shod, corn-fed and fat—valuable jacks, of course. They exchanged them for horses, mares and mules, and the stock growers of the country who purchased them thought they were getting great bargains by giving ten or a dozen mules and horses for one of these corn-fed jacks, with shoes and halter-muzzles on. But after the Missourians had got away with their horses and mules, someone whispered that these jacks were nothing but half-breeds, or perhaps El Paso Mexican jacks, which had been driven around into Missouri when young. After being pushed through on corn and fixed up for sale, they were brought to western Texas to gull the stock growers with. This did not flatter the purchasers so much, although the jacks proved valuable to them, they not having as good or better ones, and some of them none at all. They were probably half-breed jacks, raised in Missouri from Mexican jennets and ordinary American jacks or something of the kind.

Of course, after selling these animals as Kentucky or American jacks, these men would not, after practising such an imposition, come into the same section of country and do well with another lot of jacks, although there are many men here who are glad to get half-breed jacks. But they do not like to pay for full-bloods and get but half-breeds.

There are men in the country who have brought numbers of improved Durham bulls here (*of course they were Durham*) and sold them from seventy-five to a hundred dollars each at a year old. Many a dozen calves have I killed or "deaconed" at three or five days old, when the milk of the cow became good, which would have made much better animals than some I have seen that were brought here and sold, at one or two years old, for a hundred dollars cash.

Persons bringing stallions, jacks or bulls here should bring them when young, if possible. It does not make so much difference about the age of sheep, and not so much with the horse kind as with cattle. I believe that from a year or two years old is a better age than older for horned stock to bring to Texas, although they are brought here of all ages and when properly attended to, do very well. Swine, as a matter of course, should be brought when quite young. The trouble and expense of bringing a half dozen little pigs, carefully boxed up, would not be equal to that of bringing one grown hog, and nothing like as disagreeable.

Too much attention cannot be given to animals when on their way here. It matters not what kind of stock it may be. It should have the very best of attention on the way, whether brought by land or water, and when arriving here from across the Gulf, it should be taken at once to a suitable place for keeping—that is, it should be taken (hogs, perhaps, excepted) far enough back from the coast to find for it high, rolling and dry lands, good clear water and plenty of shade where it will be sure to find a good quality of grass. Imported stock generally requires more of less feeding in the dead of winter, until it become well acclimated, and even then there should always be a supply of hay or fodder laid in, in case it may be required, unless your stock should be upon a mesquite range or a plat of Bermuda. The fall or winter season is the best time to bring such animals here. By being brought in the fall or winter, the warm weather of summer comes gradually upon them, and with care they are very sure to live and to acclimate finely.

The spoiling of the cow's bag from having more milk than the calf can take is made an objection to improving the prairie cattle of Texas. But that objection amounts to nothing, if the importer has got the beef Durham breed, which are not noted for superior milking qualities. Whether they are or not, I know they will answer first rate for this country. I have traveled many hundred miles in the interior of Australia, and never in that country saw anything of the kind but improved cattle. I believe all the cattle there are of improved English breeds, and many years ago there were ten million head upon that great island. Never did I hear any complaint of the spoiling of cows' bags in that country in consequence of their giving more milk than the calf required, and of course the calf is raised there as it is in Texas—or, rather, generally speaking, nothing but the calf takes any of the milk from the mother.

I know of several extensive stock growers here (and one is our ex-lieutenant governor) who are improving from the Durham, and they make no complaint of this kind. If this is an objection, the introduction of male animals and a gradual improvement from the first to the second remove, and so on, would obviate this difficulty. Let no one give this objection to the general improvement of Texas cattle the least weight, if he has the beef Durham breed to bring here. Everybody knows that the cow of this breed is more inclined to fat than milk.

There are thousands of people in Texas who would be glad to get animals with which to improve their milch cows, or from which to raise good ones. There are different kinds that would pay well to bring here.

The milch Durham or the red milchers from Kentucky would do very well. There are, no doubt, other kinds, with which the writer is unacquainted, and some that he knows of that he cannot designate as of any particular kind, that would be first rate to bring here.

Let those who bring stock of this or any other kind to Texas, to dispose of as good or thoroughbred, bring with them vouchers from respectable men of standing in their communities that their animals are what they represent them to be and not humbugs. They can then advertise their stock in the journals of the country and offer it at public sale or otherwise. They can then, with propriety, point out to the stock growers of the country the many advantages that would result from a general improvement of their animals, and induce them to engage in the work. After having established a reputation in the business, the importer's word would be a sufficient voucher for the good blood and excellency of his animals. I believe that nothing could be done, with the same trouble and expense, that would add more to the wealth and prosperity of Texas than to turn loose upon her prairies, among her cattle throughout the state, ten or twenty thousand beef Durham, or good improved bulls—that is, if these animals were raised in the country, so that they might not die from exposure. But to raise them in the country, of course there must first be extensive importations and improvements.

It will be a glorious day for Texas when her stock of different kinds is well improved, particularly when her cattle and sheep are like the millions of these animals upon the island of Australia—when she is not outdone in the quality of her millions of useful animals. What is more beautiful for the husbandman to look upon than a large drove of entirely thoroughbred beef cattle, or a yoke of the right kind of oxen; a lot of excellent milch cows; a flock of superior mutton or a flock of fine wool sheep, heavily coated down to their toes and up to their eyes in wool; a drove of mules, fifteen hands high and well put up; or beautiful and serviceable horses, and solid, heavy and profitable hogs?

The day will come when the prairies, hills and valleys of Texas will be covered with such animals, and then the Lone Star State will be the admired of all the great American constellations. Then it is that the United States can boast of an Australia that is not across the wide deep, but near at hand, joined to her by the ties of sisterhood, in ready communication both by land and water, and all the relations that can make a great and free people a unit in feeling and interest.

In conclusion, let no man think of coming to Texas with valuable animals, unless he has a taste for the business—unless he delights in the raising and taking care of such animals. For no other kind of a man can make money at the business; and such men, I know, can do as well as anyone would wish to do at anything of the kind. Some men have no more business with this kind of work than has the "huge paw" of the honest laboring man with Blackstone. It is a business that requires constant attention, and no man but one who loves to feed and nurse his animals, whose knowledge of them is sufficient to enable him to understand their natures and instincts, whose sensibility is such that he cannot bear to see them suffer, who loves to linger around them and consider their beauty and usefulness, watch their improvement or decline and readily render any attention they may need, is fit to bring fine stock to Texas with the idea of raising from it as a business, or making money at it. Let those fellows who can lie down and sleep by their fine stock, if need be, and watch over and take care of them on their long journey, and when here give the very best attention—I say let such men or those who are able to hire such men, bring fine stock to Texas, and they are sure to make money with it, and no other kind of men can make money by importing improved animals into Texas, either to sell or raise from.

I have written this last and made many remarks that may appear as discouraging to the importation of northern or foreign animals into Texas, only from fear that my remarks in favor of the business might induce those to pitch into the thing who have no business with the like. Someone who, perhaps, without this precaution, would start out for Texas with a lot of valuable animals, whose knowledge of them might be but slight, and who might trust to others equally ignorant to take care of them, and after their dying or doing badly for the want of attention and good management, would curse the writer of this book for getting him into such a scrape.

A year or two since, I met a couple of Irishmen who were driving a fine Durham bull. They had landed in Texas just the day before. It was a warm day, and the bull had his tongue out, and although the men were on foot, they were driving the heavy fellow too fast. I asked them whose animal he was where they were taking him. They answered and added that he was brought from Kentucky. I told them that they were not in the north where that bull was raised, and unless they stopped pushing him, they would never reach their destination with him alive. I alarmed them considerably, and they afterward moved him quietly along. Had they continued hurrying this animal as they were doing when I met them, he would never have been worth anything in this country, even had they got him to his owner in the interior.

I might mention cases of caution in regard to imported horses, jacks and other animals, but trust I have said and shall say enough to the wise.

Now you that bring fine stock here, consider well what you do. With the right management and attention, there is nothing that would pay a person better; with bad management and neglect, there is nothing that would more certainly be attended with loss. Bring for a beginning, until experience shall have taught you, the beef Durham or beef kinds of cattle principally; the milk kind to a certain extent; not overgrown horses, but something like the Morgan species of the north; the improved coarse kinds of sheep to a certain extent, but the finer kinds principally. I believe what passes in the north for French merinos are the best fine wools for Texas, as they are more hardy and better protected with an oily and heavy fleece than the Saxon or extremely fine wools, and can stand our climate without shelter of any kind. As for hogs, let everyone be his own judge. I believe the big Berkshire is considered a first rate hog for Texas.

There is one thing about which, I trust, by this time the reader will agree with me, and that is that the man of limited means had better come here and purchase our common stock and improve it with animals he may bring with him or afterward get, than to invest all his means in foreign stock and run the risk of getting it safely here and acclimating it successfully—that is, if his intention is to become a Texas stock grower and not the furnisher of imported and improved animals.

I will here repeat that sheep are brought to Texas and managed with better success in the hilly regions by strangers or those who are new in the country, understanding the business elsewhere, than any other kind of imported stock, unless it is goats, a kind of stock that I have as yet said nothing of, but which, if I mistake not, is to be highly esteemed and very useful in this country.

TEXAS

Goats

To present this subject in a light that seems to me most important, I will introduce an article from the *Springfield Register* upon the Cashmere goat:

> *Some twenty years ago Dr. Davis, of South Carolina, at that time holding a government office in Turkey, actuated by a patriotism which was creditable to himself and likely to prove beneficial to his countrymen, thought of the possibility of introducing into the United States some specimens of the Cashmere goat—the celebrated animals from whose wool is manufactured the Cashmere shawls. These animals are justly prized by the people of Thibet, who reap a heavy revenue from the sale of the wool, and who had preserved the stock zealously from exportation. Dr. Davis, with an ardor equal to that attributed to Jason, in his expedition in search of the golden fleece, obtained the consent of the Sultan, and, organizing an expedition, entered the land of Cashmere, and obtained nine goats: two males and seven females.*
>
> *After great vicissitudes, and with a loss of several of his party, he succeeded in bringing these animals to Constantinople, whence he shipped them to Carolina. Six years ago the pure bloods had increase in number to thirty, besides having been extensively crossed with the native goat throughout South Carolina and the adjoining states.*

Some enterprising agriculturists of Tennessee recently organized a company, and having purchased some of the pure stock, entered into the business of raising the wool grown from the cross-breeds. We condense from a statement furnished us of this matter the following facts:

The wool from a third cross and a common goat is exceedingly fine, and is now manufactured in Scotland, where it is worth eight dollars per pound; each animal yielding three fleeces, averaging three pounds each, during the year. These animal have already been successfully reared as far north as New York, and the quality of the mixed breed wool has been decided to be not inferior to that used in the Cashmere shawls, so extensively imported into this country. It is desirable to encourage these efforts to improve the blood and the product of the animals. The production of this wool ought to be universally adopted; it does not require full blood, but the product of pure blood and the ordinary goat of the country is all that is required. It has been ascertained that the wool from goats three generations removed from the pure blood, each time crossed with the common goat, is of a most superior quality. B. F. Bristow, of Jacksonville, has some pure blood bucks, which he proposes shall be taken by the farmers of Illinois to cross with the common goats, and letters addressed to him will receive prompt attention, and be responded to with full information.

We trust our farmers will take this matter in hand and reap, at a very small cost, all the benefits to be derived from the product of a superior wool.

Now in regard to growing of this Cashmere wool in western Texas, I have to say that the common goat of the country is already quite plentiful here, and that both old and New Mexico, I should judge from accounts, are half covered with them. Although I have had several hundred of these animals in my charge in this country, and have seen a good deal of them in different parts of the world, I do not profess to know much about them. So far as sheep are concerned, I was raised with my nose at the grindstone. But goats I always looked upon as animals of little value, and for some reason or other, never had an inclination to learn much of them. But converting them into wool producers of a most valuable kind, and bringing them before the world in the light that the *Register* and others are doing, inclines me to think that they may yet become very useful animals, and that western Texas will be second to no place in the world for the production of this Cashmere wool, from which is made the finest fabrics.

The same remarks in regard to the raising of sheep will generally apply pretty well to the raising of goats, although the common goat will live and do well where the sheep will not, and will do very well in a somewhat bushy range. If heretofore they have, in many countries, been comparatively useless, they have been most at home upon comparatively useless lands—lands of sterility, whose only growth was weeds and bushes, or possibly half-naked hills with only here and there a spear of grass, a weed and scrubby bush, or briar. But when they become improved so as to produce wool worth eight dollars per pound, if possible, or even one dollar, would it be prudent to let them run among bushes and briars where they would tear, rub and pull out more or less of their valuable coat, every fiber of which, at the price per pound mentioned in the *Register*, would be an item.

It seems to me then that they would require about the same kind of range and treatment as do sheep, in which case immense portions of western Texas would be just the country for them. Should they retain, after their improvement, the nature and habit, and require the treatment of the common goat, then they certainly would nowhere do better than in this country; for I have, time and again, for years, seen thrifty and most productive flocks of goats in all parts of western Texas; and again, I have frequently seen large numbers of goats with flocks of sheep in Texas, said to be kept with the sheep for the purpose of keeping the sheep healthy. These goats seemed to be thrifty and productive, and appeared to be doing as well as though they were away from the sheep in a flock by themselves. If, then, the Cashmere goat can be successfully and profitably crossed with the common goat of the Carolinas, Tennessee and Illinois, it most certainly can be more promisingly crossed with the millions of common goats of Texas and Mexico, where they require no food, winter or summer, but what they crop from the range.

There are in Texas a few, and in Mexico thousands, of goats that are peculiar for their beauty and the value of their skins for fancy and waterproof or stormproof pantaloons. They are called by Texans "mountain goats," and they differ from the common goat of the country in their coat which is, along the back and sides, from head to tail, down to their knees and gambols, a long, flouncing fleece of hair which gives them, if they are not beautiful in symmetry, that appearance at least. Having never seen one shorn or made inquiry upon the point, I can only say that I believe them not to be of the ugly form of the common goat, but pretty in body and general appearance.

It appears to me that this mountain goat would be far better to cross with the Cashmere than the common, shorthaired ugly-looking goat of the country. If length of staple in the fleece, and beauty in general appearance of body, is an object in the improvement, then I must be right. And again, their natural proclivity to a fleece must necessarily give a greater weight and better staple to the fleece of their progeny with the Cashmere than can be given to the issue of the common, shorthaired and Cashmere goat. If, then, I am right, we have two superior breeds of goat under consideration. Shall they be brought together in western Texas, and the fleece of their issue made an article of profitable export? Are there those coming to Texas who will consider this matter, and possibly bring on with them some of the full-blood or crossbreed Cashmeres, and go into this business?—or, I might say, lead the way and experiment in the business, and demonstrate to the world that our inexhaustible range and millions of Texas and Mexican goats are to be useful in the production of Cashmere wool.

I see there is a recent importation of these animals from the Old World. There is no doubt a prospect of converting the millions of cheap Texas and Mexican goats into superior wool-producing animals.

Alpaca Sheep

Many years ago, when the writer was in New England, there was an opinion prevalent among some of the wool growers of western Massachusetts, that the South American Lima or Alpaca sheep could be successfully reared upon the Green Mountains. A number of gentlemen had it in contemplation to purchase a large tract of that chain of mountains for the purpose of raising animals upon a pretty extensive scale from the Alpaca and the large English breeds of sheep.

These gentlemen were of the opinion that the cross could be successfully made, and that the growing of the wool of these crossbreed animals would be a profitable business. That the Alpaca sheep could be raised upon the Green Mountains, they felt sufficiently assured from evidence near at hand, which was that a gentleman had a small flock of twelve or fifteen Alpacas, the most of which he had raised upon these mountains. But by careful inquiry and investigation, it became doubtful with them about successfully crossing these Limas with our large English sheep, and they gave up the idea of sending to South America for a number of the animals with which to make a start. Since then I have heard but little of these Lima sheep, the wool of a certain kind of which I have always understood is valuable for the manufacture of shawls; and if there is any part of the United States where they can be profitably raised, I should suppose that the mountains of Texas would be admirably adapted to this purpose.

I saw, in a number of *Harper's Weekly*, a year or two since, an account of a small drove of Alpacas that were driven from South America, through Mexico, to the United States, and shipped thence to England. This driving was described as a tedious and risky business. If these animals are worth all this trouble and expense to Englishmen, and can be profitably reared in eastern climates, do they not deserve the attention of Americans, and are not the mountains of Texas the place for them? I heard a gentleman say, who pretends to know, that they would do exceedingly well upon the mountains of western Texas.

Camels

Camels are already here in western Texas, breeding and doing good service for the government. And who knows how many of these ancient and hitherto, to us, valueless animals that have been so long in the hands of the Arab and aborigines, or almost exclusively controlled by people of little cultivation and enterprise, are to become a hundred fold more useful in the hands of an enlightened and great people?

[Editor's Note: It is possible that Buck encountered the animals involved in the camel experiment in antebellum Texas.

They landed at Indianola in April, 1856, and were transported that summer to San Antonio by way of Victoria before pushing further west. Matagorda, Lavaca, Refugio and Victoria Counties, as well as parts of the Hill Country are places where T. F. Buck was to be found during his time in Texas. The beasts were thriving and were set to do service in arid West Texas and beyond, but the Civil War and Reconstruction, of course, proved disastrous interruptions.]

Jacks & Jennets

I believe I have said nothing about the raising of jacks, as a business, in Texas. It is not long since I heard the representative of the district in which I am now living telling about an old gentleman who has a jennet that supports his family. He says he is well acquainted with the old gentleman, and knows him to be a truthful man. This jennet has brought him a colt every year for several years back, some of which he has sold for five hundred dollars when quite young. And from this animal, the old man raised a stock of females, besides her jack colts, which he sells only, that would produce jacks enough in a few years, if well attended, to make a person rich.

I have frequently seen in this country half-breeds from Mexican jennets and fine Kentucky jacks, that would compare well with many animals of their kind that I have seen in different parts of the world, said to be superior, well-bred animals. And I am prepared to say that almost any person in the north who has a fine jack and a little money to spare, could bring him here, and, by securing a number of good Mexican jennets at a low figure, he could get rich raising jacks—provided his management was judicious.

He would have to take some pains to acclimate his jack, and keep him in good order for the first year or two. His use to someone, over and above what he would require of him with his jennets, would pay for this and more, too. The keeping of his jennets would cost him nothing but a little watching. Neither would the keeping of his increase cost him anything, unless he preferred to grain his young jacks, which no doubt would pay in the end. If he were situated upon a good mesquite range, he would require no grain to raise half-breed jacks that would bring fine prices at from two to four years of age. I have heard of good Mexican jennets selling in Mexico for two dollars and a half per head. They can now be bought at a low figure in western Texas.

Dogs

I believe I have considered the stockraising business and the improvement of stock in Texas, with remarks upon all the different kinds of animals that are really useful and necessary in our country, unless it is some of a different species, the raising of which would not constitute a stock grower in any country. There is of the canine species a great surplus in this country, but instead of shepherd dogs, which would be useful here, this surplus is principally made up of prowling wolves, and the mischievousness of the one determines the value of the other. These shepherd dogs are brought on from the north, to a certain extent, and I am told are in great demand by many who gave gone and are going into the sheep business. I suppose there are more of them brought by land, with droves of sheep, than in any other way—which sheep, I hear, since the beginning of this work, as I anticipated would be the case, are being pretty extensively brought across the country from Illinois, Ohio, Tennessee and other states. It is almost useless, I suppose, to say to those who are coming here for the purpose of raising sheep, that they should bring with them, if possible, shepherd dogs. As game is abundant, pointers, setters and other hunting dogs are valuable here, as well.

Fowls

The raising of fowls in western Texas is the easiest of all imaginable employments. Not far from my place of writing, a farmer, during the fowl raising season, keeps his little negroes at the herding of turkeys. He raises hundreds annually and they live upon insects and grasshoppers in the prairie and about his premises. Chickens and other fowls that a person may see fit to keep do remarkably well, and can be raised in the country at little or no expense. There are fowls in the north that would be an object to get in Texas; but of the awkward Shanghai enough are here, and can be easily obtained.

General Remarks on Stockraising

I believe I have not distinctly conveyed the idea that some stock growers of Texas raise stock of all kinds, in preference to raising one kind exclusively. One I spoke of, who brands about three thousand calves, and one who raises about a thousand mules, and this is their almost exclusive business; but others I know who raise cattle, horses, mules, sheep, goats, hogs and even jacks, in moderate numbers of each kind, and also cultivates a snug field of corn, plenty of vegetables and fruits. By so doing, they can better manage to keep their stock near home, neither kind particularly interfering with the other, as the cattle range in the valleys, the sheep and goats upon the hilltops and the hogs in the bottoms or timbers. The horses generally take their choice of the cattle range, and at times mingle a good deal with them, and again the herds of horses and mules are entirely by themselves, and will have no other stock with them.

Our government has been in the habit of sending her agents to Mexico for mules. She now consumes a large amount of mules, horses and oxen raised in Texas. She does much toward establishing the price of these animals, and is apt to pay pretty well for them. Government contractors in the transportation business also consume a great many of these animals, particularly oxen.

But who are to be the great consumers of western Texas' surplus production? Her exports, of course, will be mostly north and eastward; her beef, either fresh or salt, is to be consumed by the cities along the whole Atlantic coast of the United States; her mules and horses by the planters of the south and farmers of the west; and her wool, of whatever kind, which I believe will yet be the most extensive article of export, by the manufacturers of the north and east. Then who should feel an interest in western Texas? I answer: the north, south, east and west. But if the north and east are awake to their interests, they will get and hold a controlling influence in the trade of Texas. The east should look with earnestness in this direction, for it is western Texas that will eventually be able to furnish the looms of New England with wool at about half its present cost, and leave a handsome margin for the grower.

If I am right about the ability of western Texas to raise wool at about half its present cost to eastern manufacturers, should not the east push forward and secure a steamship communication between Boston and Texas, and at the same time show to the wool growing world the great importance of Texas as a wool producing region? Dot the hills and mountains and the immense sheep ranges of western Texas with the woolly tribe, and American manufacturers will supply the place of imported goods to a great extent in the future. Why should not the east reap some of the advantages of the large and growing trade of Texas, for which so many of her articles of manufacture and export are so well adapted?

The importance of Texas and her future resources are now getting to be so well understood that she is now watched with a jealous eye by different parts of the world. Her eventual extensive communication with New Orleans by railroad will make her the principal dependence of New Orleans for her fresh beef and meat of all kinds. Who, in western Texas, would not be delighted to see a train of cars rolling away from the foot of her hills, over her plains, and through her forests, on to New Orleans, occupied by happy passengers? And freight trains loaded with beeves and veal cattle, hogs and mules, and piled several tiers or stories high with sheep, to be laid down in good time and order for the butcher or purchase? Surely such a road will someday, not far off, make the people of Texas glad, and will be the pride of and a stronghold of the Crescent City upon our trade, partiality and sympathies.

What a different country would our western Texas be if it were within a few hours' pleasant ride of New Orleans. How would the cooped-up citizens of that place rush for the mountain air of western Texas, in the dull season and heat of summer, if that road, already commenced, was pushed to completion? And how are these roads, the life of time in these days, to thread the rich mantle of western Texas and carry life and joy to the hearts of her people in the shortest possible time? Only by setting for the excellency of our country, and showing to the world its great advantages, which will call to our assistance thousands of enterprising men in the way of permanent settlers, whose interests and well-being would prompt them to labor for the cause of railroads, and whose patronage would be the needed support of such roads.

Grasses of Western Texas

Having hurried over the different kinds of stock-raising, as it is now carried on in Texas, and touched upon the general improvement of the inferior stock of the country, etc., I will now venture an opinion that nowhere can the raising of stock of any kind be made more pleasant and profitable, at present and for ages to come, than in these regions of country, and I will attempt to show the correctness of this opinion.

In the first place, there are grasses in western Texas that, in connection with the advantages of the climate, for stockraising purposes, I believe are not surpassed in the known world. The names of the most superior of these grasses are mesquite, bermuda and buffalo. The principal grass, however, of a large portion of this country is sedge. This, mixed with wire grass and other spontaneous growth along much of the coast of the country, does very well at present for the raising of the hardier kinds of stock, particularly cattle, as the business is now carried on. Further in the interior, upon the more hilly and mountainous portions of our country, this sedge grass being shorter and finer, does very well for sheep, and will continue to do so until things become more crowded here.

The day is coming when this fair land, this most favored of creation, must produce and contribute to the wealth and happiness of millions. It cannot be that a country like this, which gives forth in its rough, uncultivated state, a constant grazing livelihood to millions of useful animals, is to be always overlooked and forever left the mere playground of a few idlers, who do not appreciate the lovely spot that God has given them. This country, if generally brought into a good state of cultivated grazing, would be, to say the least, the great evergreen pasture of America—the richest grass plat in the world. I do not say that there are no industrious men in western Texas. But that there is a want of industry and of different men here to make the country what it should be, or in other words, to develop its capacities for production and usefulness, there is not a doubt.

O, that I could see western Texas, its planting portion excepted, one broad sheet of cultivated mesquite and bermuda, furnishing the consuming world more beef, wool, horses, mules etc., than any other portion of the United States of the same extent. I wish those men of energy and taste, who are looking abroad for new countries with the intention of purchasing lands for stockraising purposes of any kind, might be induced to come here and travel over this country to satisfy themselves in regard to its excellency and wonderful adaptation to this business.

But to return to the work of showing the correctness of my opinion...This mesquite grass, that I said was so fine, is of several kinds but generally known as the curly and bearded. There are large portions of southern, western and northwestern Texas that are covered with these mesquite grasses. And when I have said that they yield abundantly, and that in our coldest months, the cattle running upon a mesquite range are as fat as stock ever gets upon grass only; and that one single man with a half dozen Mexican horses can, in the way things are now managed here, attend to a thousand head of cattle; and that horses, sheep, hogs, etc., can be managed comparatively easy and cheap, I have said enough, if true, to prove that the stock business here, so far as profit and labor are concerned, is second to the same business in no part of the world. And there are plenty of good men who would vouch for the truth of what I say.

But Texas will soon be much better than now for the raising of stock. As her different kinds of stock are improved, so will her superior grasses be cultivated, and the means of keeping stock bettered. And yet, although the country is to be, someday, settled up and completely occupied, and mostly too with stock growers, its capacities for useful grazing will be manifold greater than now, and the profits of the stock business here cannot for ages decrease. At present, perhaps, nineteen-twentieths of the grasses of the country go to decay or are consumed by fire. This annual waste of grass will be gradually consumed by stock and, when too closely grazed, much of it will eventually give out and be succeeded by weeds. This seems to be the fate of the sedge grass, which suggests the idea that native vegetation, like the native man, must be pushed away, or else it must be cultivated.

I have talked of the profits of the stock business here in ages to come, and let me take a wider and far-off view of the case. The bermuda grass, I believe, is not a native of this country, but nothing grows more luxuriantly than it grows anywhere in these latitudes. There is not a mud-hole or sandhill in western Texas where it would not grow with apparent exuberance, and when set upon the rich plains or deep-soiled prairies, over rolling, hilly or mountainous lands of any soil, it loads the earth as though it were stacked by hand upon it. I wish it were in my power to describe it. Among all the vegetable productions of the earth, there is nothing more beautiful to look upon than this unequaled grass. When I think of its beauties and its many virtues,I throw down my pen to give it another scrutinizing examination and further consideration before attempting its description.

When set out and unmolested, it puts forth a fine, solid stem. From this stem other branching stems come forth, and then come clusters or cunning nests of fine spears, covered with fine blades, which, in connection with a like offspring of other grand parent stems, make up a wonderful whole of thick, verdant, sweet and nutritious foliage. It is so luxuriant and heavy that, seemingly, the earth is burdened with its weight, or in other words, having matured and given up its juicy greenness, it seems to lay in woolen warmth or shady coolness, as if to protect its Mother Earth from the effects of heat and cold, but really to fatten all healthy animals that may graze upon it. By taking several stems of this grass into the mouth and chewing it well, you will discover that it has the flavor of sheep sorrel, though nothing like as tart. My description is of new settings of this bermuda grass that have grown to maturity. If grown from old or well-formed sod, it would require a somewhat different description, and should be described in its different stages—I mean its infancy, childhood, youth and maturity. But it useless for me to try to do it justice; and I believe that words were never given to man to adequately picture the beauties of nature.

Now let us suppose it possible that, in the older settled states, or still older countries of Europe, where the lands are all fenced into lots, fields, meadows and pastures, there could be a pasture, inclosing stock, producing more grass than any land ever produced of timothy, clover or any other northern grass. Adjoining this pasture is another, which, having had no stock upon it during the fall months, has grown up a fine crop of grass and cured itself into the best hay upon the ground. Now in the midst or dead of winter, when the first mentioned pasture is likely not to afford sufficient grass for this stock, we turn the stock every day, after the dew is off in the morning, on the pasture of self-cured hay, taking it off before the falling of dew in the evening to prevent tramping and uncleanliness when wet, as well as to prevent gorging and disease.

I say suppose it is possible that such a course could be pursued, and that this much might be all the extra work required for the wintering of stock in those countries, would it not be a hundred-fold preferable and much less expensive than the present indispensable course of hay-cutting and curing, putting up for the winter? And feeding out several times a day, from four to six months of the year, besides building barns, stables, yards, racks or mangers, troughs for feeding grain, fixtures for watering, often shoveling through the deep snow to do your work, and having scores of other inconveniences that are required for the wintering or fattening of stock in these northern climates or frosty regions of any part of the world? Think of this, my cold country friends, and try to imagine the virtue and beauty of these mesquite and bermuda grasses, and this delightful country where they grow so luxuriantly...where stock live and do well, summer and winter, upon the rough, wild and inferior grasses of the country.

The buffalo grass is of short, fine growth and is not found anywhere near the coast. It is excellent for stock, where it is found in the high regions of the interior. It is very nutritious, and that portion of western Texas producing this grass is excellent for sheep and goats.

It is not upon the buffalo or buffalo grass that the writer is inclined to dwell. This animal, with the red man, is fast migrating from Texas, and surrendering the range to domesticated animals. Although there are many of them in northern and northwestern Texas, I believe they will, ere long, be among the things that once were, so far as this state is concerned. I judge from the fact that, not many years ago, they were known to occupy ranges here where now not a buffalo is seen within a hundred miles. As for the buffalo grass, it may be useful for all time to come, for aught I know. I am told by those who have traveled over it a good deal, and who have worked mules and horses upon it, that it is the strongest grass in the world, that animals will endure great hardship upon it, without grain, and that it is a good winter grass. I will venture to say that, in those latitudes where snow does not fall, the stock business can be made as easy and profitable upon buffalo grass as upon any other. At all events, I should not be afraid to risk it.

There has been a great deal said about cultivating northern grasses in Texas. This, I think, would be perfect nonsense. In the first place, there is no northern grass that would be half so good for this country, provided it would do well here, as our mesquite and bermuda. But the northern grasses, such as timothy, clover, bluegrass, etc., I believe would not do here, and so far as I know, there has been no attempt made to explain the reason why. I think I have myself discovered something to this effect, and that is that no grass can be as hardy and durable in Texas but that which is fine, compact or solid stem. No grass that puts forth a coarse hollow stalk like timothy or clover is fit for this country. Something like the mesquite or bermuda, which are next to everlasting or which are hard to kill out, will be the suitable grass for this climate when our inferior grasses give way.

The surplus and cheapness of lands in Texas inclines many of its inhabitants to believe that unless a farm is situated upon a fine stream with a superabundance of timber and water, and more than enough of everything, it is not worth having. But if people of the north and other countries understood the management of stock, or generally believed that they could successfully manage the stock of Texas, and knew the profits of the business and virtues of these lands, it would not be long before every rod of unoccupied land in the state would be made use of. But here it lies, munificently covered with grass, the great and almost only dependence of the stock grower, who rides over it day after day, and year after year, as though it was fit only for him to despise, and he calls it worthless hog-wallow, the grass of which he commits to the flames whenever it may please him—which grass accumulates for the want of stock to eat it down, the land not having more than two or three, or possibly a single head to the thousand acres upon it.

This land, were it in England or New York, would be considered the best in the world. I am now alluding to lands near the coast, but there are thousands of acres in almost any county of western Texas, which are considered nearly worthless from the simple fact that people can let their tens of thousands of cattle live upon them for nothing. What an idea! Hundreds of thousands of acres of land in the state of Texas, in the United States of America, away from all danger of Indians, in peaceable sections, the soil of which is inexhaustible, and yet almost valueless. Land that if once put in a good state of cultivation will last forever, and if well set to mesquite or bermuda grass and mowed twice a year, I believe would produce four tons of the finest hay in the world to the acre, and equally good or better pasturage. And this, too, in the finest climate, where snow is seldom or ever seen, and where cattle, horses, sheep, etc., live the year through upon grasses that are nowhere to be compared with those under consideration.

I would not have a better country than western Texas, although to those who do not look forward there are seeming objections to it now. But to the man who looks out upon the future of this great country, there is cause for wondering at the stupidity of men and contemplating coming events with pleasure and satisfaction. I said I would not have a better country than this, nor would I, for the greatest objection to it is the easy support it affords to man, thereby begetting habits of idleness. But energetic and intelligent men of the right stamp will yet fill it up. Now that the real condition and importance of the country is being understood by people of other parts of the world, it will not be long before dwellings will be near enough together upon these unoccupied lands in Texas to enable a person to see from one to the other, or to pass from one to the other in a few minutes.

General Remarks Upon Grasses

I will here mention a few facts of a general character in relation to the bermuda and mesquite grasses. I have several dozen fowls and since I have no dogs, they are shut up every night to keep them from owls and vermin that live in the timber near my house. When let out in the morning, these fowls go directly to a plot of bermuda around the house, and crop the grass as industriously for an hour's time as would a flock of sheep on being turned out of a fold upon fine grass of a morning. Fowls can nearly live upon it, if not quite.

The bermuda grass, when eaten down so close that there is not a spear, stem, blade or any part of it more than an inch long, is so thick with short green blades that there cannot be a sign of the soil seen beneath it. It is a solid sheet of short, green blades that hides the earth as well as though it were platted or knitted together. So far as this grass produces—and I believe it will, when cultivated and pastured, produce as much as any grass that grows—it is impossible for the mind to conceive anything of the kind more perfect and beautiful.

The curly mesquite is something like it in appearance, and one might suppose it to be of the same family of the bermuda. It is considered by those who are acquainted with both to be as good as the bermuda. This much is certain: in the coldest months of December through March, this grass is well known to be as good as it ever is, and I have repeatedly seen beef cattle off from mesquite ranges in these months that I believe were as good as Number One northern corn-fed beef. I do not say that they would weigh with northern stall-fed animals of the same age, but they were full fat, and their flesh firm and heavy, yet tender, juicy and sweet.

The bearded mesquite grass, besides producing a heavy bottom of fine blades, puts up a stem that heads something like barley, although in its native state nothing as heavy as good barley—yet when compared with many other kinds of grass, this bearded mesquite might justly be called grain. If well cultivated or once set in well-prepared ground, I believe it would produce seed, besides its bottom growth, what would be equal to a pretty fair crop of barley or oats.

It is contended by some that these mesquite and bermuda grasses, for fattening cattle, are equal to corn. Whether they are or not, if in the dead of winter, they will keep stock fat, they are good enough, particularly the different kinds of mesquite, which cover hundreds of thousands of acres of unsettled lands in western Texas, upon which everyone's cattle are at liberty to run.

Now, you men of money, who are seeking new and fine countries, let me ask you: How would you like to own a league of this mesquite land in the valley of the Nueces, which is as pretty valley land as there is in the world, and already covered with this wonderful grass, which lands will produce corn, potatoes, fruits and, I might say, almost every good thing abundantly—

where stock cattle can be bought at six dollars per head, broodmares for twelve and fifteen dollars per head, and where beef cattle are now worth from sixteen to eighteen dollars, and will soon bring twenty dollars per head, at three and four years old, right at home—and where half-breed mules at three years of age will readily bring from seventy-five to one hundred dollars—the lands being worth from one-half to three dollars per acre?

If the cry of gold was started across some broad ocean five times the width of the mad Atlantic, how many thousands are there who would sacrifice the little they possess, rake together every dollar they could, and blindly rush to those regions of gold—in nine cases out of ten to be disappointed, and either beg their way back or send home to get money to enable to them to return, or possibly commit suicide, in despair, for the want of luck or pluck to dig through what they thought to be a fine enterprise, never dreaming that here in Texas there are openings to which most gold diggings are no more to be compared than the least star in the firmament is to the glorious sun.

I mean what I say. There are thousands of men who went in times of excitement to California and Australia, who, had they come to Texas at the time and invested the little money that it took to get them to those countries, would be rich now and independent men; instead of what they now are—many of them not only poor, but dissipated, gambling wretches, or, what is equally bad, spendthrifts of that which by chance having made so quick and easy, they knew not its value. Casting it adrift, they soon found themselves in the rapid current of despair, and too often in the whirlpool of destruction.

Why should not men of the north and European countries come to Texas? Beef cattle are already raised here and driven to the far north, and not only consumed there but packed up and consumed perhaps by the ships' crews of every nation that may have a vessel afloat; and Texas is but just in her infancy at this business. She will yet astonish the world with her vast productions of different kinds of meat, horses, mules and wool, or anything that is raised upon grass. If a beef can be grown to three or four years of age, as some contend, for two dollars and a half; a good half-breed horse or mule of the same age for five dollars; a fine yearling sheep for a bit per head; merino wool at an actual cost of not more than six and a fourth cents per pound; hogs and other animals, fowls, etc., proportionably cheap, why will not more of that great surplus population of other countries come here and produce with their labor, energy, enterprise and capital, supplies for those whom they leave behind?

If a person owning two farms—one upon which he might live, where from four to six months of the year the earth might be covered with snow; the other farm twenty miles away, where snow might never fall; the one at home good for the growing of grains, vegetables, etc., and where labor might be cheap and plentiful for the raising and storing up of this produce; the other producing the best grass in abundance the year through, where stock of all kinds could live without any preparation for winter and where he (the farmer) could raise his stock for one-half or fourth of what it would cost him to raise it at home—would he not be likely to send his stock, with a portion of his family to attend to it, to the farm twenty miles away, and let it be used for the raising of stock, and use his home farm for the raising of grain? This grain would have to be gathered and put in order for sale or use just as much as though produced on the home farm, which we will suppose could not be done to advantage for the want of cheap and sufficient labor; and if the grass farm were used for the growing of grain, etc., it would throw the stock back on the home farm, where there would have to be a useless and laborious storing up of hay, etc., for the wintering of his stock, and six months of the most valuable growth of the grass farm thrown away, and unnecessary labor and expense used in the wintering of the stock, when it could have lived without any preparation for winter, had the grass farm been unmolested and not put to the growing of grain, or had been left to produce its winter growth of grass and constant grazing for the stock. So in regard to Texas, when considered in connection with the older states of the Union and different parts of the world.

As mankind are one great family, why should not that portion of this family at home which is yet too large, and where labor is yet too cheap, where there are many who are not needed, where the unfortunate ones of the family have too often to stint their bellies for the want of supplies or means to get them, where most of the soil could be used to advantage for the raising of grains, etc.—I ask why should not these overgrown branches of this great family send off more of their surplus labor, energy, enterprise and capital to this part of the world to produce certain supplies for those whom they leave behind, and take in exchange for them money or an equivalent thereto in such things as they might require from abroad?

Yes, since the productions of this part of the earth are so cheaply and easily brought forth with so little labor and capital; where suffering from cold is comparatively unknown and where extremes of heat are less oppressive than in cold northern climates; where there are so many leagues after leagues of cheap, unsettled or unoccupied lands, as good as can be; where is the great productive and fragrant green of the world, and beauty all around; where beggars are unknown and fortune smiles upon nearly all. Yes, I say, since all this is, let men pick up their things and come over here, and raise beef for New Orleans, Mobile, and soon for Cuba, for Chicago, New York, Philadelphia and Boston, for Canada and the Lord only knows where else. Come here and raise mules; come here and grow mutton and wool; come here and produce pork and bacon; come here and get rich out of this immense surplus of grass which is now consumed by fire upon the great pasture of western Texas. Here is the place to live; here the glittering lone star, the beacon of light to benighted Mexico, will be the great and unequaled producer for the consuming world, where millions of cattle trail the valleys and sheep are to range upon a hundred thousand hills.

The Destiny of Mexico, Etc.

I will here say that from the beginning it has not been my purpose to write anything like a thorough description of Texas or any portion of it, nor any country or portion of country. My principal object was only, in a style of my own, to give an idea of the present inviting condition of this part of the world to people of other parts of the world who may have it in contemplation to emigrate to other countries, or who, if possessed of information such as this book may possibly afford them, might be induced to look this way, and give western Texas a consideration at least, before surrendering themselves to the claims of other countries.

There is probably nothing concerning these parts at present that would be anything like as interesting to some men as the subject I am about to consider. This is Mexico, her government, people, territory, etc., and I regret that I am not possessed of information such as would enable me to handle the subject more to my own satisfaction at least. I can, however, follow these matters sufficiently for the purposes of this work.

In the first place, Mexico is what may be considered a "kingdom divided against itself," which, according to the sacred oracles, cannot stand. Everybody knows that all over this unfortunate country, there is dissension and civil war. There is no stability to her government, consequently no security of peace or property to her citizens, real or transient. Every successive usurper or mis-called president, governor or what not, every division of short-lived power, is there wringing the very hearts of a Godforsaken people, by unjust requisitions upon their property and constant arbitrary calls to arms in causes that seem to prove themselves impregnated with nothing but weakness and shame—there being no central power commanding sufficient respect and support to enable it to conquer and maintain a permanent peace.

Who can say what are the causes of this deplorable state of things in Mexico, how long they are to exist, and what will be the result of her inevitable and irretrievable fall? Does her religion, whose votaries and leading promulgators are equally important personages in the pit of the cock and bull fight, or in the house of their God, have anything to do with her shame and inability to restore and maintain order? Is the religion of the country a mere mockery, and have the long-faced and ignorant priests of this religion become, in Mexico, to be born and raised at a trade, or in a manner that debars them from a knowledge of what virtue should be or really is? Is this the cause of the trouble and inextricable difficulties there; or has this country or government sunk so low and beyond all hope of permanent restoration from the dishonesty of statesmen, politicians and military men; or is the whole country dishonest? Is the nature of the climate of Mexico, and the little exertion that is there required to live, that makes her people idle and has the sin of idleness worked her destruction? Have her mines of gold and silver been her damning curse; or is it the tendency of the Spanish race to go down, and from mixing with and extending co-equal rights to the inferior race of aborigines of the country, is it meeting, in this instance, a speedier fall than is allotted to the mother, as well as other Spanish countries? Are these any or collectively the principal causes of existing circumstances in Mexico?

Whether they are or not, if Matthew, 12:25, can be applied to her, then, according to the Bible, she or her present government is doomed to become extinct, and it may be her race of people, like the Indian, also. Is Mexico yet to be linked to the country that is characterized by the Yankee? Corruption exists more or less everywhere on the earth, but like combustible matter it will in time ignite, and everywhere burn out. Man seems to be imbued with a spirit of virtuous and intellectual elevation that at present is beyond his comprehension, and strange are the slow and tidal workings that are taking him to his distant miraculous height.

The great comprehensive truths, says President Quincy, written in letter of living light on every page of our history are these: Human happiness has no perfect security but freedom; freedom none but virtue; virtue none but knowledge; and freedom nor virtue has any vigor or immortal hope except in the principles of the Christian religion.

Is there freedom in Mexico, where any man's property is liable at any time to be demanded or confiscated by its dictators or loose government officials, or laid waste by unrestrained divisions of lawless power? Is there freedom where a man knows not when he or his property are safe, and where he is so easily condemned to death or imprisonment upon charges of crime, whether false or true; where he is arraigned by assumed authority and barbarously treated in the most rotten tribunals of justice, upon mere pretense, or for the most trifling misdemeanor? Is there virtue where a majority of men can be bought, and compelled to kill or participate in the killing of their neighbors, and where, so far as the marriage contract is concerned or vows of connubial love, affection and constancy are regarded, children know not who are their fathers?

Is there knowledge where the populace superstitiously allot to their hypocritical priests the power of invoking at their will an earthquake of destruction upon them; and is the Christian religion in Mexico? What kind of religion is that whose church levies and extorts from its ignorant supporters a tax or duty upon their infant children as they are born into the world, and when the father is unable to pay this heavenly admittance, the child is confiscated, and eventually sold by this *one and only true* church into Christian bondage, or kept as the meek or most abject slave of this sweet institution, in default of payment for the right of birth or existence?

What kind of religion is that whose priests are vile debauchees? What kind of a shepherd dog is that which steals into the fold of his own flock, and with wolfish play, throttles and sucks the life blood of its pretty and innocent lambs? What kind of angel was that beautiful little girl of twelve whom I saw decked with golden wings and sparkling diamonds, and in robes of satin richness, brilliantly bespangled, making her way through the streets, followed by those who kissed the earth whereon she walked, and she marching to the frightful ding-dong of bells, on her way to that church or cathedral, to stand as a living representative of celestial purity? And what kind of gods were they who were there to behold with lustful delight that purity and innocence, and with the unholy thought of some day robbing, with religious sanction, this beautiful creature of her most sacred boon?

And again, what kind of a god was he whom I saw in the circus of the cockfight, dictating to his whelps, and deciding the most trifling questions of dispute in regard to the game; who himself staked and won six hundred dollars on a favorite cock, after bribing the gaff-setter, and whom I saw, a day or two after, sprinkling most holy waters upon the heads of those who paid him for the pardon of their sins, and this, too, in the presence of the image of Christ and the Virgin Mary!

Can it be that anything bearing the name of religion is good and useful, and should the religion of a people, however degrading, be unmolested? "He is a bad citizen," said Napoleon, "who undermines the religious faith of his country. All may not perhaps be substantially good; but certain it is that all come in the aid of the government power, and are the essential basis of morality." Would he be a bad citizen who would undermine and scatter to the four winds the false faith of Mexico? One would suppose that the religious faith of this country might not be so good, but that it might come in aid of just such powers of government as there exist, and that it might be the basis of just such morality as characterizes her people.

Just how long Mexico will remain as she is, or how long it will be before she is regulated and protected externally as well as internally by the United States, would be difficult for anyone to say. But from the appearance of things, the time is but a little way off when something must be done in the way of consummating her connection in some way with this government. How she will be connected with or what her position will be at first under the United States, I suppose no one as yet pretends to know. But when the stars and stripes shall wave out and over her, in whatever spirit the eventful time may require, then Mr. Copperface will have to put up his knife and be still. He may jog a little at first, as does a watch after having long beat to irregular time, but when the work of regulation and protection (such as the United States would send there or extend over her) shall have commenced in good earnest, then there would be safety and security to the traveler and American resident in Mexico. Then, my countrymen, will be the time for somebody to make money as fast as by digging for gold and silver in the richest of unfailing *surely productive* mines; and who are to be the lucky fellows?

Will it be those who are living in western Texas, New Mexico and all along the frontier? Or those who are far away from the scene of action? Who are to get the best berths? Who the loaves and fishes? Who the rich claims in the mines of Mexico—those world renowned mines that the indolent people of the country have left for ages half dormant and unworked, controlled as they have been by a fast sinking, mongrel government? Said William Wirt, in his remarks upon Aaron Burr: "To youthful ardor he presents danger and glory; to ambition, rank and titles and honors; to avarice, the mines of Mexico."

You stupid fellows, away yonder, don't you see the bird fluttering? If it flies, wither will it go? Are you of those who, having eyes, see not, and ears, hear not? Here are mines that are certain to yield, and will pay for working; here is a country full of gold and silver, full of an inferior race of people, who are to be furnished by somebody's merchants and who are able and will have to work for some enlightened people—the Lord only knows who.

Full of horses, mules, sheep, goats, cattle, etc., the lands of which are not only mostly rich and fine for the culture of various kinds of grains, vegetables and fruits, "but," says a writer in *DeBow's Review*, "afford timbers for useful dyes, prized gums of commerce and medicinal extracts. Besides gold and silver, there is scarcely a known metal which is not found in her mines." Her mineral wealth must be inexhaustible. The reader can form some idea of what the country might have been, in the possession of a different people and under a different government, when he considers the amount of her church property, which is "between two hundred and fifty and three million dollars; and that at the City of Mexico, the clergy own more than half the buildings, and the whole are valued at eighty million dollars; and the income of the church is also annually eighty million dollars; the annual yield of her mines, twenty-four million dollars; her agricultural wealth, two hundred and sixty million dollars; the value of her domestic manufactories, ninety million dollars annually; her manufactories, in number, are forty-six cotton and eighty woolen, located in the central states of the so-called Republic; and about forty thousand pounds of silk is made annually at the different silkeries."

Can anyone doubt that if citizens of the United States could have known, years back, that they could be perfectly safe and could feel secure in person and property in Mexico, they would now be there in thousands, making money in some way or other? What great advantages would have arisen had such a state of things existed for the last half century or more! Why the Mexican people would have become, ere this, somewhat Americanized, and more or less fitted for the citizens of the United States. They would have become, in a measure, reconciled and to a great extent favorable to our institutions: as an evidence of which, Texas has a very favorable influence upon the bordering states of Mexico. But now nearly all remains to be done.

To take her altogether into the United States, as she is, to exercise the rights of a separate state or states, would be a dangerous experiment, and the people of the Union would probably object. To take her as we have been doing, by piecemeal and parts, would be much the safest way for the United States. This way Mexico could be regulated and civilized by degrees. By the time she should all be consumed or gradually taken up by our government, she might be pretty safely harnessed for pulling in our traces. But she has become so distressing to herself and troublesome to other countries, that this mode of disposing of her is out of the question.

She is being urged upon the United States, and if we would shape her destiny and keep her out of the hands of some other power, we must accept her in some way, and satisfy other governments in regard to her future course, or guarantee them against further injustice from her. Then, since she is to be ours, in some shape, and eventually no doubt to comprise different states of the Union, why should not all the eyes of all good men be turned upon her, and every laudable means used to meet and counteract the withering effects of her superstition, wickedness and ignorance, and fit her as much and as soon as possible for her coming place?

And how is this to be done? Why simply to turn a tide of northern emigration this way, and European emigration with it as much as possible. The western territory of the Union is already blocked in with intelligence, and is safe, and would do for a while if not settled so rapidly. The Canadas and all north of us are doing well enough, for the present, at least—for British America is free and happy. There is no danger north, east, nor west; no dreadful monster in the black garb of half-civilized superstition and ignorance to meet and overcome in those quarters; no walled-in treasures controlled by hypocrites with pretended rights and powers from on high, who can call to their support their hordes of semi-savage dupes.

The danger is this way: This very Mexico, unless wisely managed, may yet explode the great magazine of human rights. But what is to be done? The moral, social, industrial, educational, political, pecuniary and religious wealth of the United States is immense, and could pay out largely to save and elevate immoral, idle, ignorant, impolitic, bankrupt and irreligious Mexico. But the question arises, how are we to bring this healing and regenerating of things to bear upon her, preparatory to her coming into the Union, or without first taking her into the Union and subjecting ourselves to the perils of overthrow before she is put to rights and rendered harmless?

My answer is that we should take her at first as a government, subject to the United States, with an express and unqualified provision that no state should ever be formed out of her territory and admitted into the Union, until it should count a foreign white population of a certain number—a purely white foreign population from the United States and other countries, with their increase (aside from the original population and its increase), that should number, perhaps, as many as a territory now requires for admittance into the Union.

With such and other precautions, I, as an American and citizen of the United States, should not be afraid to extend our power and protection over suffering and deluded Mexico. It may be that the constitution is in the way of such a course; but whatever becomes of Mexico—whether she become British, French, Spanish, American or retains her independence—the engines of civilization should be set hard at work in this direction; and when the time or opportunity presents itself, strong settlements of the Anglo-American caste should be made in her territory among her people.

As western Texas is the center of my theme, I must not endeavor to sweep too broad a flood, and leave the whole but indifferently clean. The settling up and improvement of western Texas is itself a stepping stone at one of the doors of Mexican alleviation. Texas was herself once a broad wilderness of Mexico's, and had it not been for the colonization enterprise of the Austins, and the superior qualifications and untiring exertions of the younger Austin (details of which can be seen in the *Texas Almanac*) she might now be the victim of Mexican perfidy, and the theater of her blasting contentions. This is quite plain to Mexico and to Mexicans, many of whom are living in Texas, and constantly going from one country to the other. If, then, the force of example and bordering strength is effective and important, why not add as much as possible to the already widespread information in regard to this beautiful country? Show the profits of business, and the inviting condition of things in general in western Texas, and let the interests of men prompt them to its shores to participate in its many advantages.

While yet in sight of this Mexican question, there is one thing to which I would call the attention of all good and intelligent Americans. Missionaries are sent abroad over the earth to convert the heathen and ameliorate the condition of mankind. Is it not to our government that Americans owe their great advantages and capacities for doing so much good abroad? Is not our government the rock upon which stands the great light now shining forth to the world? To be sure, our moral and educational institutions impart to this government its vigor and the healthy blood that courses in its veins, and are its only hope in the future; and one would suppose that it might be adequate to the purpose.

But suppose seven or eight million lost people—lost to everything or wanting everything that could create or sustain such a government as ours—were to become altogether, with their country, an extreme one-side section, a part of our confederacy; might not this possibly destroy our missionary strength, and dim this our brilliant and promising light of the world, by lessening the vigor and purity of our government, unless our moral and educational strength should be proportionally increased and applied in a way to rapidly impart this wholesome blood of our nation's system into the veins of this vast but shattered additional limb of the government? If so, had not missionaries and those Americans who feel an interest in the enlightenment and saving of the world, better first see that the most attractive light, and the one by which they get their strength, and by which their feet are guided, and to which mankind seem to turn with gazing hope, is replenished and does not lose its lustrous beaconing, and secure to it, if possible, the means of shedding abroad for all time to come, before turning aside to strike their sparks in other lands, however dark and heathenish they may be?

When we consider the evils of party spirit, universal suffrage, gold discoveries and many others with which our government has already to contend, the idea of adding thereto by taking, in a body, under our shelter, even with the greatest precaution, the millions of superstitious and ignorant beings of Mexico, to be led, without principle, in elections and government matters, by their false gods, and they too (their leaders) to be led by those who might stoop the lowest or bid the highest, well may we pause and consider—well might the freemen tremble.

But, to make the thing safe, we must face the music in good time. If the thing is inevitable, "let it come," and let these shepherds, whose devotion is the conversion of sinners and salvation of souls, gather up strong flocks of intelligence and determination, and with a mild and Christian spirit make as soon as possible powerful settlements among the half-heathen of our own continent. Let someone follow in the footsteps of the immortal Austins, and colonize another Texas within the limits of Mexico.

I do not mean that I would array sect against sect, for war and contention of any kind are repugnant to my feelings. I simply mean that I would bring the intelligent and true Christian in friendly contact with the deluded and ignorant; I would bring industry, enterprise and everything that is great and good in contact with idleness, degeneracy and everything that is detestable and wicked. In short, I would gentle and improve the savage and degenerate blood of Mexico by establishing colonies or strong settlements of the Anglo-American race among her people, in every advantageous quarter that should offer money-making inducements, or honest paying labor and available operations, such as tilling the soil, digging for gold and silver, lead and minerals of any and all kinds, merchandising, raising horses and mules, growing wool, manufacturing, mechanism, school teaching, medical practice, etc. And how, I ask, could good and Christian men better serve the dearest causes than lead the way in such enterprises, for the purpose of opening new channels and widening the fields of true Christianity and human happiness? And lastly, but by no means leastly, for the purpose of forging new chains with which to bind the north and the south forever together and render, if possible, this imperishable Union.

To be understood, I must go a little further on before returning to my dear western Texas. I would not disguise the fact that I would have northern people go to Mexico, and so would I like to see southern people go there. This mingling and mixing up of the people of the different sections of our great country will, no doubt, prove a most excellent thing. But the advantages of populating Mexico with northern people particularly would be immeasurably great. Settling down in the south seems to create in the northern man a higher estimation of the Union, a renewed love of the Union, a deeper, stronger Union feeling and determination to stick to the Union. It gives an attraction to the center. Northern people in the south, so far as I know, do not meddle with or disturb its institutions—I believe they generally become southerners, according to the correct definition of the term, so far as it should relate to sectionalism in these United States. But they have a desire to see the difficulties of the country peaceably settled, for the idea of being politically separated from the land of their birth, if I am correct, is perfectly distressing to them; therefore, here in Texas or Mexico, if it belonged to the United States, the northern man would impress upon the minds of his children that there is a great North for which he has a peculiar regard—a great and powerful people in the north, whose intelligence, energy and advancement are unsurpassed, and with whom we should strive to cultivate the best political feeling and relations; for, in connection with them, we are the greatest, the most powerful and happy people in the world; but separated from them, wherever the line might be drawn, we would be comparatively weak, and we know not what might be our destiny as a people—separated, we could not look forward with that certainty of a glorious future for our country.

Such and like sentiments instilled into the minds of their children could not fail to impress them with a deep sense of the importance of the Union, and create in them a true attachment to it, and they would cling to the land of their fathers and strive in support of that which should promise the most—the undivided Union—which would counteract, to a great extent, the feelings imbibed by those who are raised to hate the Union on account of sectional agitation. It may be urged against this position that California would willingly cut loose from the Union, but I don't believe that she could be induced to do so, even though the mountains should ever divide us and no railroad should ever connect us. But, with a connecting railroad, nothing short of a Rocky Mountain earthquake that should throw us all into the seas, would ever separate the Pacific from the Atlantic states; and should such a thing occur, the separation of California from the Union, and the South from the North or the Union, are by no means analogous cases.

Whoever goes to Mexico should strive to elevate the character of the Mexican people, that they may reach the day, as soon as possible, when they will not be liable to dangerous corruption in elections and governmental matters—what an immense work of ages, and how deserving the particular and even the undivided attention of the good and wise men!

Before returning more closely to the idea suggested at the beginning of this section, I will say that I do not pretend that the introduction of a different religion in Mexico could by itself work out the desired effect and save the country. But it is industry, commerce, internal improvements, mechanical, educational and useful operations of a thousand kinds, in connection with good morals and truly religious operations that would work the desired change.

For a Protestant minister to go to Mexico alone and divulge his doctrines would, I suppose, be equivalent of throwing himself away. A physician, of Protestant faith, by himself in Mexico, if he were so disposed, could be a stronger champion in a religious way than could a Protestant minister, provided he should perform miraculous cures or ingenious humbugs. Mexicans would, of course, then consider him inspired, and were he equally accomplished in religious matters, he could show them the wonderful works of Protestantism. I mention this merely to show the character of the people, not to encourage such work. If a man, to accomplish good and drive out corruption, must resort to the scurvy pretense of Catholicism as it exists in Mexico, as has a certain *distinguished character*, there might be danger of his becoming corrupted also, or it might not be an unnatural conclusion to suppose him possessed of but little good to impart to others. For selfish ends such a course frequently works admirably.

Settlements in Mexico by people from the states, of a reasonable religious faith, should they be successful in their business operations and rapidly improve the country that they might possess, would naturally suggest to the mind of the Mexican that a better religious faith than his might be requisite to prosperity and advancement with him or his countrymen. I mean to be understood that to show the Mexican the effects of a good religion is better than to undertake to force a different religion from that of his own upon him; for when urged, in a religious way, he would hug all the closer to the one he now professes, though it might curse him ten times as much as it does.

To make successful settlements in Mexico, the direction of the work in its beginning and for some time, aside from religious matters, should be entrusted to the right kind of men who are already acquainted with the language, the people and the ways of the country. The simplest, unpretending, yet wide awake means will accomplish the greatest good in the business. To get together a company of outcasts, whose only idea is gain either by fair or foul means, without a shadow of creditable aim in life, without a claim to respectability, and lost even to self respect, and make a great cry and empty show and pretensions, without work, good regulation and behavior, is all nonsense. No enterprise ever so promising could amount to anything in the hands of such men.

If a settlement for mining purposes is to be undertaken in Mexico or any of these half-savage frontier territories, the managers of the enterprise should know what they are about—they should not only be men of great determination and skilled in the business, but they should know beforehand that they are to be very sure of successful digging in some well-known locality, where they are equally certain in regard to the durability of water around these diggings, or that their preparations during a dry season are where they will be followed by a sufficiently wet one for their purposes. They should know about the means of getting their supplies, and what would be their probable cost up to the time that they could probably realize from their operations, and prepare for the same, otherwise they would be liable to discouraging disasters that might break up the company and possibly the enterprise altogether.

The men who direct such enterprises in Mexico, or upon the frontier, should encourage the emigration of stock growers and farmers to settle around them if possible. There are sections in the vicinity of gold and silver regions in the frontier territories that are well adapted to the raising of sheep and goats in particular.

New mining companies should determine from the start to have no cut-throat, gambling, dissipated, disturbing wretches among them; and I believe every such company, that is sometimes from necessity shut out from society and away from the gentle influence of woman, would be much more successful if they would not only interest moral and Christian men in their enterprise, but a minister of the Gospel, who should hold forth in a religious way occasionally, and use his influence to keep up as good a state of morals as possible. Good industrious men who are willing to work for a fortune, if rightly organized in companies and directed by the right kind of experienced managers, could certainly do remarkably well at mining and stockraising in various quarters, along the frontier of Mexico, and the day is most certainly not far away when Mexico itself will be the place for many an American to fill his pockets in a thousand different ways.

I am credibly informed that at this time, good broodmares can be bought in the interior of Mexico at five dollars per head, but men who have recently been there for the purpose of bringing out horses have returned without them, considering it unsafe to purchase even at these low rates; for, as they say, could we have gotten them for nothing, there was no certainty of our getting them out of the country at this time, and there would have been a risk of losing our lives with our horses had we brought and started with a drove. The troubles of the country endanger a person's life as well as property, and he knows not on whom to depend.

But when it is safe to go there again, which time is not far off, then, my boys, whom it may concern, think of buying good broodmares for five dollars per head, and horses, mules and sheep in proportion. Think of this country (western Texas) where, if you please, you can bring them and so profitably raise mules, horses, produce mutton, wool, etc. Of course in times past, there has been more safety in going to Mexico, and thousands of men have been there and made their piles at buying stock for a mere nothing; and when there is safety in going there again for this purpose, there will be plenty of men at the business, some of who are now standing ready and will pitch in as soon as a good time offers.

Some will buy for the purpose of driving to Texas and further north to dispose of at wonderful profits, but many will buy for the purpose of raising stock in Texas; and I know of men and am credibly informed that there are plenty of such who are anxiously awaiting the day when a large portion of Mexico, at least, will be ours, or when they can go with the protection of our government in some way, and purchase and peaceably occupy the most beautiful spots in Mexico that they ever saw; as fine cultivating lands and stock ranges of entire mesquite, in as delightful a country and climate as there is anywhere upon the face of the earth. That will be a glorious day for this part of the world! Many a Yankee will find his way into Mexico, and buy out the indolent and ignorant Copperface for one-fourth the value of his stock and lands, or I might say for mere nothing; i.e. one-fourth the value to the Mexicans. Yes, that will be a glorious day when the Yankee boys shall plant an influence in fallen Mexico that will live and grow into such importance as could originate with no other people.

The people of Mexico are a mixture of Indian, Negro and Spanish, and not a few full-blood Indians and negroes are her lawful citizens and participate in the affairs of her government, religion, etc. Should not this be a caution to abolitionists? The people of poor benighted Mexico are inextricable in their connection with the red and black man. Is not their destiny Indian-like? Then if Mexico, from her connection with the wild man and other causes, must go down, why not the Yankee boys occupy her lands and country, whose resources and importance she has not developed?

I would not be understood that the present inhabitants or population of Mexico should be murdered or driven out of their country, but I would be understood that the present peculiar race of Mexican people are doomed to become extinct, and that sooner or later their country is to be characterized by a different people, although fragments of its present race may remain there for many centuries to come, but will all eventually be absorbed in another race, or lose its identity in some way and the country take an entirely different character. And why not that character be given by the Yankees? Why not let this country or this part of North America be characterized as is the whole United States? Cross the ocean and drop your anchor in any distant port of the world and go ashore, and when the stars and stripes are seen waving out upon the pennant of your ship, if every stranger in your presence, when speaking of that ship don't call her a Yankee, "I'll pay the drinks." I care not from what port of the United States she hails, or where first launched, whether California, Norfolk or Portland, if she hoists the flag of stars and stripes, when the skipper of any vessel of any nation levels his glass upon her, the chances are his words will be, "She's a Yankee."

Go abroad, my friends, born and raised wherever you may have been in these United States, and you'll be called or considered a Yankee. And why should not the people of the United States be termed Yankees? Is that word only an empty sound? Does the word Yankee associate with it nothing but the idea of Indian jargon or wooden nutmegs and the like? Who settled the Yankee states, and how have these states progressed? What are their institutions of learning, and what have they been for many years? What of their common school system? Who are the statesmen and learned men they have produced? And what of the talent, the capital and kind of men they have sent abroad and settled all over the United States? What of their manufacturing and mechanical skill and productions? What of their inventive genius? What of the part they act in the commercial world, and what of their religious influence? Who and of what are the commercial cities of the United States composed? Who built the canvas city upon America's ocean? Who built the puffing city upon the inland waters of the United States? Trace, if you please, the great men and the great works of the whole United States to their origin and see what the Yankee states have done—see if they have been behind. Why the very father of Texas first saw light in old Connecticut.

One would suppose that the writer might himself be a Yankee, but he is not, nor is he so bigoted as to suppose that there is no people or doings of consequence in the United States but Yankees and their doings. He is, however, somewhat acquainted with the Yankee character, and although a little partial to them, he would like to see coming to western Texas with them people from all the states and all parts of the civilized world. The mines and wealth of Mexico, her vast extent of rich mesquite and cultivating lands, and her millions of stock are near to our borders, and will be easy of access to the western Texian, and the man who is on hand in time will be the lucky fellow.

So come from everywhere; this mixing and crossing of the white civilized races of mankind is, no doubt, developing great advantages in the Yankee nation, but with the Indian and Negro a dangerous experiment if Mexico is an example. Yet it might not be so dangerous for the descendants of Great Britain and their adopted kin, or mixed offspring of the United States to live and intermingle with the inhabitants of Mexico, since the Anglo-American's destiny seems to be ever upward and onward. Yes, it matters not what obstacles are in the way, what difficulties to overcome, what seeming impossibilities to do away and heights of trouble to surmount, wherever you may find him, he is ever moving steadily but surely on and upward—mid sunshine or tempest he is alike the same, immovable in his lofty course, and it seems that his is to be the universal language of the northwestern world at least. The English tongue seems not to know the world fail—it carries with it universal success.

Land, Crops, Dwellings, Water, Yellow Fever, Northers, Etc.

Although I have talked of most of these matters in connection with stockraising and other subjects, yet I have considered them with only here and there a word, and will therefore take them up separately, with a few remarks.

There are millions of acres along the coast of western Texas, and some distance back from the gulf or bay shores that are known as "hog-wallow lands." Why they are called by this uncouth name, I do not know, unless it is from the peculiar unevenness of their surface, which someone may have supposed to look like the work of hogs. This unevenness resembles, in appearance, land in the north or timbered countries, which are sometimes left, after chopping them off, for the stumps to rot out and take to grass, with their natural and unmoved surface, which is often very knolly, or, in other words, a surface of little hillocks all covered with grass. The soil of these hog-wallow lands is generally black, and when moist, of a sticky nature. Its depth is from two to six feet underlaid with a kind of clay.

A person might travel a lifetime over these lands and never see a stone of any kind. It is heavy, at first, to break up, but after being well turned over to a depth of three or four inches, the grass and roots of all kinds in it readily die out, after which one or two good plowings make it pleasant land to work, if worked at the right time. It should never be tramped by stock of any kind when cultivated.

It is very productive, and it is said will never wear out or become less productive from use. These lands are disagreeable to work when wet but if, after a crop of corn or anything else is taken from them, they are soon well turned over in a dry time, they require no further preparation for the next year's crop; and they are generally very pleasant to plow up directly after taking off a crop, before the soil gets too much settled or too closely packed together. I have known men to fail to make a crop by putting off the plowing of their hog-wallow field until late, when it would break up in lumps, and having no rain in time to pulverize and settle the soil together, the seed could not take root for want of moisture. There is an unusual amount of lime in this soil, and when plowed in a dry time, it turns up more or less in lumps, and becomes dryer, and not until a soaking rain will it settle and prepare itself for a crop. Water slackens it nicely together as it does lumps of lime.

With the right kind of management, a person is very sure of a crop upon these lands. For, although when stirred up, it admits the air and loses its moisture, when pulverized and settled with a soaking rain, it excludes the air and retains its moisture better, perhaps, than almost any other soil. The farmer who is ready for the spring season is almost sure of a crop upon these or almost any other lands in western Texas. Corn can be planted as early as January and as late as April, for a summer crop, which gives ample time for replanting, should the first planting fail from any cause. In fact, men have been known here to raise two crops of corn upon the same land in one season.

The hog-wallows, generally speaking, are among the cheapest lands in the state. This is owing, to a great extent, to their being heavy to work, and their want of timber; but as I have said, one great reason of their being so cheap is that men can pasture their tens of thousands of cattle upon them for nothing, and this is the case with much of the land throughout the state—the principal owners possessing each so much that they are unable to fence it, and many of them having no stock to inclose, would have no use for their lands were they able and disposed to fence them, as, if fenced, they could only be available to them as pasturage; and furthermore, it would not add enough to their value and quicken their sale sufficiently to pay the landholder for fencing, unless a similar course should be generally adopted throughout the state. Under the present state of things, this would be an impossibility. Could landholders prevent the stock of the country from running upon their land without fencing it up, the land might sell for much more than it does, and at paying rates. But there is no remedy, and consequently the bulk of the range of western Texas is to be open to the world, as mere commons, for many years to come.

Although the lands along the rivers and principal watercourses in western Texas will be generally taken up and occupied at no very distant period, the great body of prairie range, out from and between these watercourses, will be open perhaps for ages to come; and to this, as much as to almost any other one thing contained in these pages, would the writer call the particular attention of the reader. These lands, out from the streams and timber, in the center of our broad prairies, many of which are first rate, can generally be had at a much lower figure than can those bordering on the streams and having timber upon them; though, Lord knows that well-watered and good-timbered lands are for sale here without end, and cheap enough as yet.

But if I were this day a newcomer in western Texas, and here for the purpose of raising cattle, horses, mules and the like, knowing what I do about these open prairies, I would by no means give up my undertaking should I find every rod of land having water and timber upon it to be taken up or occupied. I would, in this case, go into the center of these prairies without water or timber, either upon hog-wallow or rolling uplands, and establish a ranch. If in the hog-wallow prairie, my ranch could be upon the highest, sandy piece of ground I could find—plenty of which sandy places, a little above the common level, are always to be found. Here I would have one advantage, at least, over my neighbors upon the watercourses, and that would be a fresh start of grass in the greatest abundance, which could not be as abundant as where cattle had been grazing for years along the streams of water.

I would go to work and supply my cheap place with its deficiencies—water, shade and wood—and then I would haunt my cattle or stock to it, which of course would necessarily be Texas stock at the beginning. I would supply my want of water for stock purposes with ponds, made of natural basins, to hold water as long as possible, and with wells, whose depth in hog-wallow lands need be only from fifteen to thirty feet. For the use of the house I would have cistern water, the best in the world. To supply my want of shade, I would inclose a small piece and plant China trees, the rapid growth of which would not only soon supply an abundance of shade, but plenty of firewood and posts for fencing, which latter would be equal to the best of cedar.

I should work all the time with the satisfaction of knowing that my lands, which cost me a mere trifle to start, could not be beat for the growing of the better grasses, should I ever see fit or have occasion to cultivate them; that they could not be beat for the growing of sugar cane; as good as any in the world for the culture of the grape; second to none for different kinds of fruits, such as peaches, figs, pomegranates, etc.; second to none for nearly every garden vegetable that grows; good as need be for corn, the great feeding staple of the country. I say, as I should work all the time with the satisfaction of knowing all this, and that the breeze from every quarter would ever generously cool and relieve my healthy home, and that railroads would surely someday add much to the value of my property and to the cheerfulness and happiness of society and the country around me, and that people would, after a while, wonder that they had not seen and appreciated the value of these open prairies, I could work with good cheer and always be pleased with my location, and never envy my neighbors upon the watercourses, whose lands may have been, from the first, supplied with timber.

These remarks I design as a hint in regard to the future of these broad hog-wallow prairies, some of which, in places, are twenty and thirty miles from river to river, with but little timber or watercourses of any account to be seen. There is, now and then, a scattering cluster of trees to be seen, when crossing almost any of these open prairies, and plentiful ponds of good stock water, unless it is in a dry season.

Considering the cheapness and virtue of much of the land in these prairies, the great abundance of grass around many fine localities that could be selected, they are deserving the attention of stock growers who would secure a cheap home and an extensive range. Such lands can be had for a half-dollar per acre, in considerable quantities, which sooner or later are sure to be very valuable. But at present, as I remarked, there is no end to the fine land for sale in western Texas, that is well-watered and timbered, and upon durable streams. Should I select a ranch or permanent home for the stock business in an open, rolling country, deficient in wood and water, I should pursue pretty much the same course as in the open hog-wallow prairies.

When I talk of these open, central points in the broad prairies of western Texas as places of residence, I am looking some ways into the future. But it is a subject that is every day gaining importance, and attracting attention in this country, particularly the purity of the air and the healthfulness of these dry localities. Although we can, in certain places and directions, travel twenty or thirty miles in open prairies, from one durable river to another, without seeing much of any timber or water, a person cannot, as a general thing, travel these prairies, in any direction, that distance or half that distance without crossing rivers, creeks, lakes and bayous, the banks of which are more or less heavily timbered; some enough only to furnish firewood and partial conveniences to those who may settle upon them; others affording the greatest abundance for fencing and all conveniences of the kind to the settler, except a good quality of lumber. Plenty of Florida pine, however, and sometimes northern pine and spruce is to be found in our bay towns suitable for building purposes.

The sweet potato is the most natural for the climate and soil of western Texas. Irish potatoes grow well but they do not seem to keep from one planting to another, which is a great objection to them here. But the sweet potato, which I believe is relished by all, after eating them a while, produces abundantly, and there is not difficulty in storing them in a way to keep. Irish potatoes are raised a good deal in the spring, and serve until sweet potatoes come on in August. With few pains, the Texan can have fresh Irish potatoes the year through; and this pain is only to raise a spring and fall crop and, if it is necessary, to lay them by or store them a while for planting purposes. They need only to be carefully spread in a dry loft or packed in perfectly dry sand. I believe they are sometimes left in the ground from one planting to another, when two crops a year are raised.

Nearly all the garden vegetables of the north grow finely here. Spring and fall gardens are always made, and many persons contend that the fall is the best season for gardening. Any time in the winter, unless it is an unusual one, a person can go into our well-managed gardens and take from the ground, just as they grow, beautiful vegetables of almost any kind. Finer melons never grew than can be raised in any part of western Texas, and two crops a year at that; also pumpkins or anything of the kind produce finely.

The most natural fruits for this country are peaches, figs, grapes, plums, pomegranates, strawberries, raspberries, tomatoes, etc. Apples, pears, quinces, currants and the like, are produced in the interior upon the hilly and mountainous portions of western Texas. Oranges, dates, bananas, etc., are raised, but not generally, and with what success I am unable to say. The following are some of the wild fruits that are common in parts of western Texas: wild plum, wild grapes, haws, persimmons, mulberry, dewberry, blackberry, etc.

Nearly all the bottom forests and much of the upland timbers of the country are more or less burdened with wild grapevines, which are loaded with grapes, and they are the most thrifty and juicy grape that ever grew. Europeans are here manufacturing wine and brandy from the grapes of these broad vineyards—not of forbidden fruit, but fruit to which the whole world is welcome. These grapes are too tart for pleasant eating, though some people take them down with a wonderful relish, as though they were half sugar. When fully ripe, the black, luscious-looking fellows are very tempting. They are excellent for jelly, jam and preserves, but require a little more sugar than the cultivated grape—I should suppose about the amount that is required for currants. Any amount of vinegar, of the best quality, can be cheaply and easily made from this immense forest production. I have drank excellent wine from this abundant fruit. The cultivated grape, of any kind, can be and is, to a certain extent, successfully engrafted into the wild vine. The fruit of the wild vine is called the mustang. To make up the one-hundredth part of this spontaneous production of western Texas into wines, etc., would give employment to hundreds of people; and when the labor and expense, attending the establishing and cultivation of vineyards, is taken into consideration, one would suppose that the making up of these free grapes into wines, brandies and the different relishes, for which they are good, might be made a profitable and extensive business.

When at Cape Town, Africa, the writer saw the free use of wine, which was largely produced, but was not comparable, in quality, to an article made here in western Texas from the wild or mustang grape. Of course, I do not allude to the Constantia wine of Cape Good Hope, but to a common article called Cape Town wine. Is it not reasonable to suppose that a country producing the wild grape so vigorously and abundantly, would be surpassing fine for the choice kinds? The truth is, this heavy, black hog-wallow land is as good as the best in the world for the raising of grapes, thousands of acres of which, all around where I am now writing, are to be had for the small sum of from fifty cents to two dollars per acre.

The haw, such as is found in western Texas, is delicious eating. I have eaten haws from trees in this country that I would prefer to the raisin. I have often wondered what they would amount to if cultivated. I have found them in the winter, in thickets of brush, plump and sweet—not as rich as the raisin but better by far for anything like sumptuous eating. They hang on the trees nearly all winter, and ripen by degrees. The frost sweetens them, unless after freezing they are too much exposed to the sun and wind—and I believe too much exposure to frost injures them. The best that I have found were protected in thickets of brush from the effects of the frost, and from the sun and wind, which, after freezing, dries them up and makes them tasteless. From the fact that very thrifty trees grow the choicest fruit, I am inclined to think that the fruit would be much better if the tree was cultivated. The color of the haw that I alluded to is blue, but I believe it is generally called the black haw.

The dewberry is very much like the blackberry of the north. The vine, however, is different, being low and scrubby. There is generally plenty of this fruit almost anywhere in western Texas. It ripens in the spring and lasts but a short time.

The mulberry affords beautiful fruit, in abundance, either wild or cultivated. There are several kinds found in our forests.

The river bottoms or valley lands of western Texas are generally of a deep, dark soil and are very productive. They are occupied extensively by cotton and sugar planters. The forest growth of these bottoms is live oak, pecan, elm, prickly ash, hackberry, mulberry, white oak, pin oak, wild peach, etc., etc. Of these, the live oak, wild peach and others are evergreen. The wild peach, mulberry, magnolia, China, etc., are fancy trees, and are growing about many of our dwellings, ornamenting and shading our dooryards. Our most lasting forest timbers are live oak and mulberry. Timbers for millwright, wheelwright and other mechanical purposes are plentiful here.

The crop of the pecan tree puts many an easy and honest dollar into the pockets of many Texans. The annual export of pecans from western Texas is very large. Early in the fall, hundreds and thousands of Texans resort to the river bottoms and timbers to gather pecans.

As for flowers, this of all others is nature's favored spot. From the remotest mountain top to the seashore, some gardens are richly robed with all the mingled colors and beauties of the rainbow, and so pleasant to the smell and refreshing to the eye that words fail to convey an adequate idea of their loveliness.

After leaving the black hog-wallow lands, which are mostly covered with sedge grass, we generally find a good deal of dark sandy soil and a rolling country, and besides the timbers upon the rivers and creeks, there are immense forests of open timber or timbered prairies, well supplied with the sedge grass and watered with beautiful streams. The principal timber of these forest prairies are the post oak and black jack, interspersed with a variety of other kinds.

Here and there among these post oak lands are rich valleys, particularly along the rivers and creeks. There is now and then a piece of hog-wallow, but long stretches of entirely post oak or timbered prairie lands of a sandy soil are found, which are not as rich and productive as the hog-wallow or bottom lands—but let the reader bear in mind that these are all good grazing lands, and that when it is required, they will all produce the mesquite and bermuda grasses most plentifully. Nothing will grow more luxuriantly and produce more abundantly upon any land, in any part of the world, than will these grasses upon almost any of the lands of western Texas. These leagues upon leagues of sandy post oak lands, although now apparently idle, are to be the pride of the country as vast sheep walks and general stock growing regions.

The mesquite lands or those lands producing the mesquite tree and grasses have a dark rich soil, of good depth, and are generally of a smooth surface. They produce fine corn, potatoes and vegetables generally, and in the interior or mountainous regions fruits in abundance, and I am told the different kinds of grain of northern growth. The principal native tree upon these lands is the mesquite, and I should say it is almost the only tree found upon ranges of entirely mesquite grass. This tree produces a bean which Mexicans feed to their horses, for which I am told they are excellent. The mesquite is generally a scrubby tree, but grows sufficiently tall and straight to make fine posts, which are said to be as lasting as the cedar of the north. The roots of this tree afford an excellent article of dye, and I believe are an article of export from Brazos Santiago. The root is said to comprise much the larger portion of the tree.

The Mexicans and many Americans build from this timber picket fences. These are made by digging a narrow ditch the width and depth of the blade of a spade, in which the bodies and limbs of trees, after being chopped the right length, are set on end as close as possible to each other, around which the dirt from the ditch is closely packed, there being no frost to heave a fence or foundation in this country. When built of mesquite, this makes a lasting and profitable fence, and such fence when in order is very safe, as nothing will attempt to jump it when of ordinary height, say from four to five feet. I have seen a fence of this kind made of live oak trunks and limbs, which I believe may last fifty years and more by sometimes resetting.

The Texians have another way of building a good fence with this scrubby timber, with which the settler would soon become acquainted. It is made by throwing the dirt from a ditch all on to one side, in the line of a fence, and planting the short pickets on the embankment. The ditch and the elevated foundation serve as a good share of the fence, as pickets of three or three and a half feet in length will do in this case.

Post and board fence is made to a great extent in this country, half the height being frequently made up as above, of a ditch from which the dirt is thrown along one side, around a row of posts that have been slightly set and which are deeply and firmly secured in their place by the embankment thrown up from the ditch, which may be two and a half feet in width and depth or less. Two and sometimes three six-inch fence boards nailed on to this row of posts, upon the bank of the ditch, is all sufficient for the purposes of the country. Sometimes when lumber is hard to get, a ditch answers the purpose of a fence; again, large fields and pastures near the coast, where lumber is nearer at hand, are fenced with posts and boards altogether.

A little back from the coast, stone fence is sometimes made, though seldom. Concrete fences are being made pretty extensively about the towns, for dooryards, gardens, etc. Rail fence of various descriptions is made, but I do not see any here that will compare with the capped and ridered fence of the north—I mean such as are capped with inch and a half stuff, bored with three and four-inch auger, and this too with horse, steam or water power. Hedging is being pretty extensively commenced, of which I shall speak further on.

The mesquite timber is excellent firewood, and besides affording posts, pickets and the like for fencing, it is used for other conveniences by those who are settled near it.

The man of intelligence and refinement, who has cultivated a taste for agricultural pursuit, and can appreciate things in general pertaining to the business, and who is capable of the least sublimity of thought, will, when traveling over fine mesquite ranges, after considering their beauty and many virtues together with the excellent climate in which they are, turn his thoughts upon the Creator with a bursting soul of purest thanks for the feelings of delight and enjoyment that here possess him. In fact, anywhere in western Texas, such a man with a knowledge of what is around him, would feel that the Founder of the universe had been partial to and had wonderfully favored this beautiful nook of creation.

The highlands of western Texas are more of less gravelly and rocky and, as I have before said, produce finer quality native grass than do the low or coast lands. But the reader will not forget that each and every section of our country has its peculiar natural advantages over other sections. Large ranges of these high, rolling and hilly lands, with hardly a tree upon them for miles around, except now and then a single tree or cluster, are well supplied with clear running streams, having branches which find their way between the hills down to that stream whose bed may be sand, pebble-stone or rock.

When upon these high but gentle hills, which slope in every direction, valleys are all below and around you, and you are seldom out of sight of timber. Again, these beautiful hilly prairies are nicely studded with live oaks and thinly scattered with a variety of trees, single and in groups, which gives great relief to the eye—there are also splendid water-holes, found in some sections, which are beautifully overhung with live oak or shaded in their loveliness by a collection of trees and bushes of some kind. It is upon these countless hills the soul may expand, the thought take wing and fly away into ceaseless space, to return laden with gems of beauty. Here it is that man feels that the Creator, in every star and figure of beauty, has pointed to happiness; yet, poor man, he forgets it and goes again to grovel in the dust! O that I might forever remain in those transports of purity and happiness experienced upon these fixed billows of earth! O that I might forever forget every sinful thought and always contemplate the beauties of nature in this, our lovely land!

Kind reader, is there anything about your present abode that is distasteful or offensive? Does the sun of your life seem checkered, darkened or obscured? If you would cut loose from every bondage of mind and be contented and free, come and dwell upon the highlands of western Texas. Plant upon the mount of your choice a seed, root or something for every hair on your head, and rear around you golden trees and nature's beauties of our own fair clime. Remember that the turf of your lands, if not at first, may be made of bermuda and mesquite; and that if your cattle do not range upon a hundred hills, the cupola of your dwelling with the aid of your glass may overlook the herds of useful animals scattered upon ten times a hundred hills. This is no delusion, no foolish absurdity, but the simple truth. It would be very difficult to give the reader anything like a correct idea of all the variety of appearance in the vast surface of rolling country in western Texas.

We find many of the rolling, open prairies well-skirted with timber, in every direction, where there are most lovely sites for dwellings and general improvements, with excellent lands for cultivation. The world of timbered prairie lands of western Texas is generally rolling and hilly, but often level. Fine streams make their way through these timbers upon which there are settlers and plenty of room for more. Stock is generally found in such ranges by hunting up and down the watercourses and in the open prairies. Here, men raise their field of corn, and have their hundreds of hogs about them which feed and fatten upon the mast of the timbers. Wheat, oats and the like are here produced, but so far as I know, northern Texas and the mountainous portions of western Texas are better for northern grains than are the rolling lands midway between the mountain ranges and the coast of our country, yet I have seen good wheat grown within forty miles of the gulf coast.

People are settled here in these timbered prairies from Pennsylvania and different northern states, and are doing well with their few cattle, well haunted—a bunch of mares kept together by a herder and generally having a bell-mare among them to which every horse of the herd comes attached—and besides a fine lot of hogs many of them have a flock of sheep. One acre in fifty of these lands, however, are not occupied, and they are generally selling very low. There are small rich prairies, free from timber, scattered all through these prairie forests, varying in size from a hundred to a thousand acres each, and more. Of course towns or trading posts are convenient throughout the settled portions of western Texas.

Of the extreme back and entirely mountainous portions of western Texas, the writer is unable to say much. From accounts one would suppose them to be the finest of country, producing a great variety of fruits, vegetables of every kind, wheat and all northern grains next to perfection, and unequaled for the raising of sheep, and surpassing fine for all kinds of stock. Almost anywhere in the interior of western Texas there can be found timber convenient to any place upon which a person would be likely to settle, and but a few miles from anyplace he could possibly pitch upon, which timber would answer the purpose of building, unless a person might be pretty independent and could afford to buy lumber.

This is a country where, I believe, aristocracy does not, as yet, kill people—although there are those here who feel their importance. There are men of wealth and respectability, in the country, who are living in comfortable log and picket houses, and who have their frame dwellings yet to build, when they feel that they are well prepared to do so. A person can do more as he pleases here, free from the ridicule and gossip of others, than he can in some countries. He can put his money into stock and build a picket or log house, living in which he can be comfortable and respected. After getting well underway in the stock business, of any kind, he can do almost anything he wishes. There is nothing like having a lot of beeves, mules, wool, mutton, fat hogs, or the like, to turn off when money is needed, especially when it costs but little or nothing to raise them. And again, it does not require an airtight or closely built house here to make a person comfortable. I have known newcomers to bring stock from northern and eastern Texas into our western prairies, and live with their families for a year or two in strong cloth tents, until they could inclose a field and make it convenient to built a house, which might then be of pickets, logs or lumber, according to circumstances. The stock business in this country, if well looked to, will enable the stock grower to erect a dwelling and make such improvements as will suit him, and have at his command such luxuries as the country affords.

The material for dwellings and all kinds of buildings, aside from the bottom or upland timbers, some of which split well and make good shingles, is by no means very limited or inferior. Besides the extensive lumber yards in our coast towns, the owners of which are always prepared to furnish the most extensive builder with lumber, sash, blinds and everything of the kind complete, we have in sections inexhaustible quarries of stone, some of which is soft and easy to work into a wall, and after years of exposure becomes extremely hard, and I suppose, everlasting. Most of the building rock of the country is easy to work, and is considered first rate for the purpose. An abundance of lime rock is found in sections, and much more lime would be consumed in the country if the men were here who would construct kilns and supply the demand, which would certainly be moneymaking, as the stone and wood in places would be, if not quite, almost free.

A vast deal of building is now going on in different parts of the country, the principal material of which is concrete, and I discover that almost anyone of any ingenuity can build up a wall of this material. The wall is carried up with a box or boards, between which layers of the concrete, in the shape of mortar, is packed and left until sufficiently dry, when the boards are elevated for another tier of mortar, and so on until the wall is completed. There are various ways in which this material is made. In the interior I have seen good fence and buildings constructed simply of lime and pebblestone, or gravel. If the wall made of this material is finished with a coat of cement upon the outside, I am told it makes a durable and I can say a very pretty fence. There are fine concrete dwellings in the country made from shell, the lime of shell and cement, mixed into a mortar and laid up, as above, with boards, and afterward plastered with cement. Chimneys are also built in this way for frame houses, and with common brick and other material.

I should have said, when speaking of building rock or stone quarries, that in places stone houses are much in use, and continue to be built. Clay bricks are, in places, made and some of our finest buildings are of this material.

Rain water is used to a great extent in this country, and consequently cisterns are required. In the coast towns there is actual necessity for rain water, but the interior is abundantly supplied with cool springs and clear, running water; yet, from a belief that has become quite prevalent here, that rain water is the best in any country, many will have cisterns although they may have good wells and fine running water just before their doors. These cisterns are, to a certain extent, made of wood in the usual old fashioned way, and set outside upon the ground, or sometimes under shelter of some kind, or in a room designed for the purpose; but by the permanent settler they are mostly constructed underground, of cement and shell, pebble-stone or gravel and cement, or else of brick and cement. It requires some experience to build a cistern of this kind, in a way not to crack. There are many striking proofs of this healthfulness of cistern water in other countries than this, and after drinking it a while, most persons prefer it and will use no other if cistern water is to be had.

It is superfluous to add that the climate of western Texas is delightful. I think in Martin's history of Australia, it is said that to go from Europe and live upon this island is to renew one's youth. This, I think, may be as justly said of western Texas. I believe this climate has as beneficial effects upon the system of the foreigner or newcomer as the climate of Australia or any other country has upon its emigrants.

Certain it is that the climate has a tendency to work out all the impurities of the settler's system; and if he is a little careful in his acclimation, to assist nature a trifle, by taking simple alternatives occasionally, he can lay a new foundation for fine health and a good old age without any difficulty. Of course, a stranger seldom thinks of doing anything to cleanse his blood or give a healthy action to his liver or system, whose pores, secretions or general internal action, gets clogged in throwing off the overcharge of thick blood, animal heat, etc., that he brings here, and must be got rid of to prepare the system for uniform good order and healthy action in this climate. Consequently, his acclimation is more severe and of longer duration.

What I mean by this is that men who come here from cold countries, whose systems, of course, are to undergo a change, unless they take some simple alternative in the spring of the year or before the heat of summer to assist nature in effecting this change, may be troubled with boils, prickly heat, or something of the kind upon the skin, which sometimes annoy the foreigner a short time in the heat of summer for two or three seasons, particularly those who go back from the coast and work for a living. But as acclimating never kills or injures, and as it is but a wholesome renewal of the system and drives all corruption, aches and pains away, what objection is there to it, since after its experience, one is in a way to enjoy uninterrupted good health, buoyant spirits, or as much so, probably, as he would in any other part of the world?

Some people never experience any difficulty from acclimation, and it may be that a majority of foreigners never feel the least inconvenience from it; but as I am not writing a one-sided thing, I have thought it best to consider this as an objection to the country, though it is not really an objection, but an evil, so that good may come.

It is no doubt the impression of many that yellow fever is a common thing in western Texas, but this is not the case. It is only here when brought by vessels from New Orleans or some other seat of the disease; and when here, it is confined to the bay or coast towns to which it is brought, unless those who have been exposed to it go into the interior before attacked, and then the healthy air of the country will not suffer it to spread or take the form of an epidemic. A person living five miles, or even half that distance, away from these towns, apprehends no danger from it. It has been taken a short distance into an interior town of eastern Texas, but it is so seldom an epidemic here, and then only when brought by ships, that I do not know why this should be a place for its consideration, any further than to say that if those who come to western Texas by the way of New Orleans will come when this fever is not an epidemic there, they will meet with no danger from it, and never have any fear of it here, unless they should see fit to remain in one of two or three particular towns of the coast where it might be brought. Of course much of the emigration to western Texas is by land through northern Texas. Such emigrants would never hear of yellow fever any more than where they now are. When we are relieved of the vast amount of vegetable decay in western Texas, no part of the world will surpass it as a healthy region.

The changes of weather in winter in this country, or the sudden appearance of *northers*, when the weather is extremely warm, would be a little surprising to the foreigner, and one would suppose that those northers might be very injurious to health, but I am not aware that they are particularly so. I have no doubt, though, that people from neglect or carelessness, or from accidentally being out without a blanket or warm clothes to put on when these northers come up, sometimes engender disease from the effects of the sudden change.

Although they are frequently pretty severely felt here, they would, in northern states, be nothing more than snug, bracing, fall weather. A person, sometimes, has due notice of the coming of northers, but at other times they will take him by surprise. In winter, snow-caps or huge white clouds are seen in the heavens—dark, towering clouds in the north, with vivid flashes of lightening—and you are in a perspiration from the still, sultry warmth of the weather. You may conclude that a norther is brewing, and ere long a distant roar and rumbling, as if all the mad things of earth were let loose together, tells you that hot and cold elements have met and that *mucho frio* is the hero of the battle. The chilly warrior has overwhelmed his foe and comes down upon him with all his fury and devilish howling.

To experience one of these very sharp fellows on a broad prairie in the night is indeed a grand but dismal event. The howl and fury of the winds as they meet you, the dark dungeon gloom that suddenly envelops the whole heavens and earth, surprise and bewilder you. But he that has braved the terrific smothering snow-seeps of New England or the Canadas would laugh at the norther's frolics and liken him to a roaring but harmless bull without horns.

The beauty of the scene follows the dark and frightful beginning as the dark clouds subside and leave the heavens in the clear, starry light, except the frown upon its northern brow, coupled with its dreary, icy face. As you travel along and look out upon that chain of prairie fire, with its swelling and receding links, and think of yonder comfortable home, the whistling winds and grizzly north lose their terror, and you are inclined to think it is not so bad after all.

These northers contrast wonderfully with the hot *siroccos* of Australia. When traveling with a party of miners in that country, we were once overtaken by one of these hot winds. We at once put out our cart-horse, which seemed to care nothing about the heat, and we made for trees and places of protection before the winds attained their greatest heat and severity. To try the affects of the hot air, which we had been assured would not injure anyone, we would occasionally jump out into the full blast of wind, and could hardly help thinking that our whiskers were snapping with fire, and that we would roast to a crisp in a few minutes then back to shelter we would go. But what astonished us most was seeing those who were acclimated to those winds, traveling along in them as though they were a luxury or nothing to be dreaded. It was supposed at the time that these hot winds were from some unknown desert, somewhere in the unexplored regions of the interior.

All the coast country of western Texas is usually favored with a delightful sea breeze, penetrating from thirty to forty miles into the interior, and there is generally a fine breeze to relieve the settler or traveler in the open prairies of the interior. The heat upon these prairies is seldom as oppressive as one might suppose it to be. As for cold weather in western Texas, many of these pages were written in a cluster of live oaks, in the open air, in the dead of winter; and so far as cold is concerned, a person could write most of the year, exposed to the elements, with only a shelter overhead to keep off the occasional rains and the sun. Wagoners and thousands of people sleep out the year through in western Texas. With two good soldier-blankets, a person can lay out the coldest and wettest nights that we have in these regions.

Rivers, Towns, Grasshoppers, Etc.

The largest river in western Texas is the Rio Grande, the boundary between Texas and Mexico, its length being some eighteen hundred miles. It is navigable by steamboats several hundred miles from its mouth. Much of the land upon this river is beautiful for cultivation, but it is generally cultivated by irrigation. A majority of the inhabitants along this river are Mexicans, and except Brownsville and the towns near the mouth of the river, there are none but Mexican towns upon it. There are American merchants and trading characters among these people along the river, and also American stock growers. There are many forts on the river which makes the country safe for American settlers.

By irrigation, they raise almost anything upon these valleys—corn, wheat and various grains, a variety of fruits and vegetables, sugar cane, etc., are here grown to the greatest perfection, in the peculiar way of farming in droughty countries, and that portion of western Texas along the Rio Grande and some distance east of it may be considered a droughty region. But even with this objection, it is considered an excellent stock country. It is generally of mesquite grass, and its want of water for stock purposes can be easily supplied in the way pastures are sometimes supplied with water in Kentucky, which is by plowing and scraping out basins, in which the stock grower feeds his hogs a while for the purpose of packing and filling the pores of the earth to prevent the basins from leaking, after which, when once filled, it affords a permanent drinking place for cattle.

A similar course could be most successfully adopted in this droughty section of western Texas, where durable rivers, creeks or lasting water are wanting, and many advantages would arise from this natural want of water, upon the principle that "there is no great loss without some small gain." In the first place, any amount of these lands can be bought for a mere nothing, which are as good as can be, only wanting water in dry times; and by supplying this want as above described, the stock grower could control his cattle with but little trouble and hardly any expense, as they could not leave their only place of drinking unless in a wet time, when they would not go far away on account of being so thoroughly haunted in the range of these artificial reservoirs. Natural basins can be found all over this country, which, with a little shaping and fixing, after having hogs or stock of any kind fed, penned or yarded, as the northern man would say, in or upon them, for a while, would make durable watering places for stock of all kinds.

The droughty country to which I allude is principally between the rivers Rio Grande and Nueces—a vast mesquite section, a majority of which is yet to be occupied, and where beautiful tracts of land can be had for a trifling expense—where as fat beeves and stock ever grew upon grass can be found in the dead of winter, and where charming homes and fine stock farms can be made and improved in a way to be forever valuable and profitable. All of this particular section of country is near Mexico, and he who settles it will have but a short trip to make over into the mongrel nation and pick up an honest dollar in the way of trade.

The outlets of this section of country are Brazos Santiago and Corpus Christi. There is steamship communication between Brazos Santiago, Brownsville (or the mouth of the Rio Grande) and New Orleans, and vessels from different parts of the world visit Brazos Santiago with cargoes of merchandise for the Mexican trade, and the trade of the country generally in the vicinity of the Rio Grande. These vessels are here loaded with such articles of export as the interior of Mexico and the Rio Grande country afford; some of which are hides, lead, wool, bullion, etc. The steamships, besides taking merchandise, bullion and usual articles of export, carry away to New Orleans sheep, cattle, goats and horses.

Certain gentlemen have imported into the Rio Grande section of western Texas, improved animals, by steamship from New Orleans to Brazos Santiago. Such animals being for the improvement of their Texas or Mexican stock, it might pay to look into this matter and see if more of such animals are not needed in this quarter. I have seen large flocks of sheep that were raised on the Texas side of the Rio Grande, and it is no doubt a fine sheep country and a good stock country generally. There has been a profitable and extensive trade carried on with the Mexicans at Brownsville, and other points on the Rio Grande, since the Mexican War, and fine fortunes have been made and are being made in this trade by merchants and tradesmen in those ports. There have been several attempts made to carry trade into Mexico by unlawful means (filibustering) which is a proof of its being profitable.

I will state that, as this matter is about going to the press, I hear of the disturbance of Cortina and his followers on the Rio Grande, but I can confidently say that this disturbance will be of short duration and all will soon be quiet. The state of Texas has taken measures to protect its own frontier, and it will hereafter be perfectly safe.

Corpus Christi, the other outlet to this droughty country, is near the mouth of the Nueces River on Corpus Christi Bay. It is a place of no very great importance yet, and it may never be large; but a town of no little importance must eventually grow up somewhere not far from it, where vessels of good size will make a safe and convenient landing. The trade of the whole Nueces valley and country for some distance in the interior will support Corpus Christi or some other town that may grow up near the mouth of the Nueces River, which valley and country I have already said is a beautiful mesquite region and is unsurpassed for the stock business. To travel up the Nueces River in the valley, upon either side, is what the husbandman or farmer of any country would love to do, and by doing so he could judge something of the capacity of western Texas for the raising of stock, and also of the necessity of a good outlet seaward for this and adjoining country midway between Port Cavallo and Brazos Santiago.

There are several little coast towns between these two points that claim to be good seaports, and some of them will no doubt be a place of importance, as the country around and the interior that would support it is at this time considered the tip-top cattle range of western Texas. A good seaport and an attractive point of trade somewhere as near as the mouth of the Nueces as possible is deserving the attention of merchants and men of commerce. Corpus Christi and small towns along the coast between it and Powder Horn have a small craft communication with Powder Horn. But the people of these towns look forward to the day when they will have a direct communication with New Orleans and other important places of trade.

In regard to this droughty section, I would not be understood that the Nueces River country is an unusually dry section. It is along the Rio Grande and the country east of it toward the Nueces that is considered droughty, but it is to this same section of country where I hear that men have taken their small stocks of cattle for the purpose of being better able to control them than they were in ranges where water is more plentiful.

That we have dry seasons in western Texas no one can deny, and this is the case in Australia and parts of California, and I believe all extensive stock countries where snow does not fall. If western Texas were a wet country, I do not believe it would be a good stock country, and it, no doubt, would not be as healthy a country as it now is.

I have noticed particularly that stock of all kinds get fatter and do better in dry seasons than in wet ones, although they have to walk much further for their water. I have also noticed that stock of all kinds are freer from disease in dry seasons. In wet seasons, sheep, unless rightly managed, contract the liver rot and other diseases that in dry seasons there is not the least danger of their being troubled with. In wet seasons, horses are troubled with Spanish fever and other diseases that are hardly known in dry seasons; and a winter following a very wet summer and fall is sometimes very hard upon stock, particularly if they get thin and weak, when they frequently bog down in their places of drinking.

But wet seasons are not frequent and the death of an animal in Texas is no calamity to speak of—it costs so little to raise them that a stock grower hardly notices or speaks of the loss of a few animals. Cattle and horses die here upon a prairie and their owners hardly ever take the pains to know the cause of their death. It is but recently that people have been at all particular about skinning and saving the hides of animals. Diseases, however, among stock of any kind are very few in this country, and the loss of stock from diseases in western Texas is small compared with such losses in northern states and other countries.

In regard to the effects of drought upon crops, the writer can say that western Texas truly and strictly speaking is a natural stockraising or grass country, although corn and other grains, by being early planted and well attended, do very well in all parts of western Texas. I believe that almost everything that grows can be produced here; yet, owing to the somewhat uncertainty of the seasons and other causes, western Texas is considered a better stock than a general farming country, although some men say there is no better farming country.

Portions of western Texas have been troubled with grasshoppers, and to my certain knowledge, small sections of country have been stripped of their last spear of vegetation by these insects. Such has been the case in Australia, and still wool, beef, wheat and everything as usual was produced in the country, except in certain small sections. Grasshoppers are liable to visit any country, but, I suppose, are more likely to trouble open prairie countries than those generally occupied, fenced up and cultivated. I have noticed that tramping and cultivating wild lands drives away mosquitoes, and I have no doubt it has a tendency to keep away grasshoppers. It is a well known fact that there is a peculiar insect destroyer for almost everything that grows out of the earth, and that these insects trouble different parts of a country at different times. Why grasshoppers should visit western Texas again more than any other country, I cannot see. Their flight when last here was northerly, and where they will take up, lay their eggs, hatch and destroy again, nobody knows.

This talk about dry weather and grasshoppers in western Texas may cause the emigrant to pause for a moment, but if stockraising is his object, let him come on, for drought or no drought, hoppers or no hoppers, nowhere upon the face of the earth can this country be beat at the stock business. I sometimes think that those sections of western Texas where dry weather prevails most were designed purposely for sheep—sheep seem to do remarkably well in our high and dry regions. Some of the finest sheep ranges in Australia are equally as dry as the droughty sections of western Texas. In conclusion upon the subject, the weather in this country, as a general thing, is about as the stock grower is glad to be favored with, on account of ticks, branding their increase, diseases of stock and other reasons that I will not use space to dwell upon. For my part, I prefer pleasant dry to disagreeable wet weather.

The Colorado, which I shall consider as the eastern boundary of western Texas, running somewhat centrally through the state, is the next river east in size to the Rio Grande, and is about seven hundred miles in length. Were it not for contemplated railroads this, no doubt, would eventually be made a navigable river for some distance into the interior. The Colorado bottoms are of a rich, reddish soil, and are generally pretty heavily timbered as far up the river as the rich bottoms and planting lands extend. The width of the Colorado valley is from four to ten miles. Sometimes upon one side, the naked prairie bluffs on the river and, upon the opposite side, the valley may be six miles, more or less, in width and heavily timbered. Again it is heavily timbered and is a wide valley upon either side of the river.

This river is settled by cotton and sugar planters, and most excellent crops they do raise. Much of these bottom lands are yet unoccupied and are for sale, but at higher prices than are the prairie lands adjoining them, or the lands generally from that on to the Rio Grande, excepting now and then rich bottom lands upon smaller rivers. Large stock growers are also settled upon the Colorado, and their cattle range over the vast prairies upon whichever side of the river the stock grower may be.

The river empties into Matagorda Bay, not far from the town of Matagorda. Small craft ply between this town and the main landing of steamships (Lower Indianola) in the bay. Persons wishing to settle west of the Colorado River generally land at Powder Horn or Lower Indianola, which are one and the same. The reason for confining myself to this particular scope of country is because the most of it is never troubled with snow, and a vast amount of it is covered with mesquite grasses, and is otherwise so favored that I believe no extent of country of its dimensions upon the face of the earth can excel it for stockraising purposes. Of course I do not include the desolate plain above and between the headwaters of the Colorado and Rio Pecos.

The next river of any note west of the Colorado is the Navidad, which connects with the Lavaca River a short distance above its mouth or entrance into Lavaca Bay. Texana, a town of considerable trade and importance, is situated on the west bank of the Lavaca River. It is the trading point for stock growers and planters for many miles around. Small vessels ply between this town and Lavaca, an extensive port of Matagorda Bay, about ten miles from the principal landing of steamships in this bay.

Austin, the capital of the state, is on the Colorado, about two hundred miles from its mouth. The rich bottom lands of the Colorado continue on above Austin, and the rolling prairie lands about this city correspond with much of the beautiful country, and would give the traveler a pretty good idea of large regions of western Texas. I am sorry to say that I am able to take the reader above the city of Austin with only hearsay and what I have read. From the many representations that I have seen and heard in regard to the country west and northwest of Austin, I can safely say that it is a charming sheep country. But when you say north of Austin, you are getting into regions where the winters are more severe, and where I suppose it is necessary to store up hay, etc., for the wintering of stock. It may be, and I am inclined to think is the case, that with the cultivated mesquite and bermuda grasses, rightly managed, northern Texas could winter stock generally without storing up hay or food to any great extent. That northern Texas is a beautiful stock country, all who are acquainted with it agree. But I wish to confine myself principally to a section of country where stock can live upon grass the year through, without having hay or anything else laid up to feed in the winter. This section comprises most of the immense country between the Colorado and Rio Grande Rivers, over much of which the writer has repeatedly traveled.

The largest rivers between the Nueces and Navidad are the Guadalupe and San Antonio. These rivers connect some distance above their entrance into Espirito Santo Bay, which bay has a navigable connection with Matagorda Bay. The navigation of the Guadalupe has been several times attempted, and steamboats are now, irregularly, running as high as Victoria upon this river, from the landing of steamships at Lower Indianola, but without much hope of successfully continuing.

The San Antonio, and several of the rivers of westers Texas, are frequently spoken of as affording good water privileges for mills and manufacturing establishments requiring any amount of power.

Cotton is extensively produced along the Guadalupe, and its valley, for some distance, is wide and heavily timbered. There are also several very pretty towns situated on this stream. Cotton is produced to a small extent on the San Antonio, but I believe this river is considered the terminus of cotton lands westward and southward in Texas. I am so little acquainted with the culture of this great staple of the south that I should be out of place in attempting to say much of it. But I can say that cotton growers in Texas seem to be wealthy and prosperous people, and, so far as I know, a democratic people, of whom a country should be proud.

The timbered valley lands of the San Antonio are principally below Goliad. Above this town the river is not so heavily timbered. But its valleys are, to an extent, of beautiful mesquite lands on to the head of the river. The country through which the upper San Antonio runs may be considered a pretty good mesquite section, though its grass is considerably mixed with sedge.

There are several pretty towns on the San Antonio River. Goliad and San Antonio are the two largest, and the last mentioned is the second largest city in the state. This city is about a hundred and sixty miles from Matagorda Bay, and a railroad from Lavaca and Lower Indianola—several miles of which are already completed, and thirty or forty graded—will soon, no doubt give San Antonio and intermediate stations, and the surrounding country, a quick communication with the gulf towns and New Orleans.

Besides the rivers already mentioned which empty into the bays along the gulf coast, between the Rio Grande and Colorado, there are numerous others, the most of which should, perhaps, be denominated as creeks. There are also bayous of various sizes, between these rivers, making out from the bays, which reach but a short distance inland. The outer coast of this section of country is almost a continuous narrow strip of land, with only here and there a pass or an entrance from the Gulf of Mexico into the different bays, which are a continuous body or connection of waters along the whole coast of western Texas. These narrow divisions of land or islands between the gulf and bays are generally stocked with cattle.

The rivers which I have mentioned and many others of less size, are of course never-failing, and when the usual places of watering sometimes give out, cattle and horses resort to these rivers for water. There are also many streams, scores of which are of good size, making into these various rivers, and this portion of the state, except for the droughty section alluded to, is well supplied with running water.

The hilly or mountainous portions of this section of country are more of less favored with beautiful springs of fine cool water. There are also many small lakes and ponds of various sizes, which are fine for the watering of stock, and many lagoons or places of water in the beds of creeks that cease to run in dry times—such as half the island of Australia depends upon to a great extent.

Now it is to these many rivers, bayous, lakes, ponds, creeks, lagoons etc., from the Colorado to the Rio Grande, and from the gulf coast of this country to the headwaters of the Nueces, San Antonio, Guadalupe and Colorado Rivers that I would call the attention of settlers. I have already spoken of the droughty sections; of wide prairies wanting in wood and water; where good lands, in any quantity, in the finest of ranges for stock, can be had at present for a mere nothing; and have shown how easily and advantageously these wants can be supplied. But now it is to the tens and hundreds of thousands of acres of unoccupied lands along and about these many streams of water, that I would call the attention of those who come to western Texas to engage in the stock business.

This large tract of country, of several hundred miles square, embracing a great variety of soil, surface, general appearance and climate, will every foot of it produce, in abundance, the fine grasses of which I have said so much. And unless its climate materially changes—unless the sun withhold its present warmth, or some unforeseen evil befall it—this particular section of country, when clothed in rich, cultivated grasses, will be a second Eden, producing as much for the support and good of mankind as any section of country of its dimension in the wide world.

He that gets a home here will be a fortunate person. For here, he or his children, with the least exertion, could never want. To point out the thousands of desirable, unoccupied localities and fine tracts of land fronting upon rivers and durable streams, or those that are plentifully watered and timbered, or in the midst of good ranges, that would do for stock farms, small and large, in this section of country, would be the work of years upon a thousand pages. But before concluding this work, I will give the reader some directions about finding desirable localities and cheap lands for stock-raising purposes in western Texas.

FISH, OYSTERS, SHRIMPS, TURTLES, ETC.

The rivers, creeks and lakes of the country are full of fish, and its bays, bayous and coast afford the finest of oysters, shrimps, turtles and fish. The principal freshwater fish of western Texas are the cat, buffalo, sucker, trout, pike, perch and carp. The jew, red, mullet, flounder, sheepshead, trout and bass are some of our great variety of coast or saltwater fish. The various ways of taking fish, oysters, etc., in this country are the same, I suppose, as in all countries. The turtles to which I allude are as heavy, when alive, as our beef cattle ordinarily dress. They are caught in different ways, but the easiest way of getting them is to surprise them when out of water, or when benumbed with cold on the beach, and whap them over their backs, where they will lay powerless. Alligators are plentiful in the bayous, lakes and large ponds of the country, and they are in all our rivers that empty into the gulf, but principally low down in or near the mouths of these rivers. Their oil and skins are said to be valuable.

Game Animals

Hunted in western Texas are deer, antelope, mountain sheep, wild goat, buffalo, wild cattle, wild hog, javelina or Mexican hog, bear, wolf, prairie dog, panther, leopard, wildcat, tiger, coon, opossum, jack and common rabbit, squirrel, otter, beaver, muskrat, etc.

Deer—He is no huntsman at all who can't go out and get a fine buck or doe in a short time, in any prairie of western Texas. A slide or stone boat, drawn by a yoke of oxen, is a good thing upon which to secret oneself for the purpose of approaching deer.

The usual way of hunting these animals is on horseback. The huntsman, to get near to a deer or herd of deer, gets down and walks by the side of his horse, which is frequently trained to the work. Deer are easily approached by fixing the horns of these animals upon the head, with a dress and action in imitation of their own.

When traveling through our prairies in the spring of the year, young fawn are often started from their hiding places, and are easily run down with a good horse. The venison of this country is most excellent eating, and there is hardly a family in the country who does not use it to some extent, and many families are seldom without it. Deer are here seen in droves of all sizes; bunches of ten, twenty and thirty are very common, and a hundred are frequently seen together; and I have been in plain sight of a thousand and more at once, not far from the Nueces River, but in different droves.

Antelope—The antelope is much the same, at a distance, as a deer. They outrun the deer, and are noted for their beautiful appearance when running. Their horns are short and nothing to be compared with those of the deer. I believe they are seldom seen along the coast, but are as plentiful back among the hills and fine sheep ranges as are deer upon or near the coast. The meat of the antelope is most excellent, and they are as easily taken in the interior as deer are upon our coast prairies. I am told that by the swinging over the head a piece of red cloth attached to the end of a ramrod, antelope will circle around and come within shooting distance of you. I was recently told by a wagonmaster that the men of his train shot not less than thirty of these animals while coming through upper western Texas, and that he saw thousands of them. Their size is about the same as that of the deer, but I believe they are somewhat more compact and heavy.

Mountain Sheep—The mountain sheep I have never seen, but have repeatedly talked with those who have seen them. Some contend that what is called the Rocky Mountain sheep is really the antelope. Whatever they may be, they are really a curiosity. That they will throw themselves from a great height and light upon their skulls and wonderful horns, I have been assured is a fact by those who have taken pains to know. That the concussion from the fall or leap, in the way that they strike, does not kill them, is not so remarkable as is the fact of their ability to throw themselves such a frightful distance and control their position in the air, in a manner to strike upon their horns in the peculiar way that they do.

WILD GOAT—The wild goat I know nothing of except by hearsay, unless they are the same as the mountain goat to which I have alluded. Judging from descriptions I have heard of them, I should suppose them to be the same, which leaves the supposition that they may be domesticated.

BUFFALO OR BISON—The buffalo or bison I have already spoken of as migrating back from the settled portions of western Texas. They are to be found in upper western Texas, and are hunted by gangs of Mexicans, Indians and white men, who jerk and save their meat to some extent, and preserve their hides for robes, etc.

WILD CATTLE—Wild cattle, besides those to which I have alluded, are found in places where there are no other cattle. They are hunted in places principally for their hides, but their meat, tallow and horns are saved to a great extent. The Rio Grande country affords a vast amount of beautiful hides. Wild cattle will soon be killed off or driven beyond the limits of good ranges for stock in western Texas.

Wild Hogs—Wild hogs are plentiful in this country, and many a happy day is there spent in the bottoms and timbers or western Texas in pursuit of these animals. Dogs are trained expressly for tracking and baying hogs, which enables the rifleman to approach and send his bullets among them. Some families in western Texas have but little hog meat except what is secured in this way. But hogs are frequently caught in this way which are not considered wild hogs. They may be those having the earmark of the owner, which from long use in the bottoms and timbers have become so shy that baying them with dogs and shooting them is the most expeditious way of getting hold of them just when they are wanted, which is during cold snaps of weather when their meat can be safely cured.

Javelina or Mexican Hog—The javelina or Mexican hog is an animal that I have heard much of but have never happened to see, although I have been in and around those mesquite thickets, the Brazados of western Texas, where they are said to be plentiful. These animals are a singular species of hog. They are small, but when molested are very fierce, and it is a dangerous business for man or dog to meddle with them without using great precaution in the attack. They are said to have a kind of musk-bag upon their body, which makes them offensive to the smell, and unless it is taken away as soon as killed, they are considered by some as not fit to eat. Others say that they have eaten them regardless of this and doubt their having any such offensive part about them. They are sometimes caught and sent abroad, and sell for a good price as a curiosity.

Bears—I suppose there are several kinds of bears in this country, but I have only seen the black bear, which is quite plentiful in parts of western Texas. There are many men here who delight in hunting the bear and who have good dogs for this purpose. Bear meat in this country is fine eating when fat. They are somewhat destructive to hogs in the bottoms.

Wolves—There is no wild animal in the country so destructive as the wolf. They are of two kinds—prairie and mountain, or coyote and lobo. The howl of these devilish things is yet to be heard upon almost any prairie in western Texas. They are, however, easily killed by scattering pieces of meat around the prairie, or wherever they come, with a little strychnine carefully secreted within the meat. Stock growers sometimes drag a dead animal or a large piece of meat some distance through the prairie, and occasionally leave a piece of meat properly dosed with poison in its trail. Wolves, if there are any about, will follow up this trail and eat the poisoned meat, and are generally found dead not far from where they find these effective doses.

Stock boys, when hunting cattle, frequently start up these rogues and give them chase. A number of us once came upon a wolf in the act of eating a calf that it had just killed. We took after it but it was so full of blood and fresh meat that it would occasionally stop and try to throw up to relieve itself and make better headway, but we were so close upon its heels he could find time only to gag and be off. The boys were soon throwing their ropes, at which he would snap and dodge; but a rope was finally fastened upon him. The horse dashed away with one end of it at the horn of the saddle and the other snugly around the forefoot of the wolf. He dangled and dragged along until he seemed quite willing to lie down and die from the blow of a quoit that was dealt upon his pate.

These animals do no destroy as many calves upon our prairies as one would suppose they might. Whenever a young calf is caught by one of them, it gives the bawl of alarm and all the cattle within hearing rally around, make a great fuss and drive the wolf away. I have seen the prairie wolf take hold of a young calf when there was no other cattle but the mother of the calf about, and she alone would compel the wolf to let loose his hold. I only once saw an affray of this kind where, after the cow had driven the wolf away from her calf, the whelp jumped at the cow and for a whole hour, at intervals, would cut up his antics around the mother for the purpose of getting the calf a little away from her in order that he might take another hold. But the calf instinctively kept close to the mother while she was desperately thrusting her horns at the cunning wolf.

Sheep and goats, unless in an old settled place, require constant guarding in some way against these animals in western Texas. Many wool growers in the upper country have proof folds for their sheep, and others have their shepherds lie down at night near their flocks with their dogs about them.

Prairie Dogs—Prairie dogs are sometimes brought in from the interior of Texas and domesticated. They are a curiosity, and being small, many people are glad to get hold of them as pets. Since the discovery of the

Pike's Peak mines, I see much is said of them by the papers throughout the country; but I have seen no account of their flesh being good to eat, which I hear is the case.

PANTHER, ETC.—The panther, leopard, wildcat and tiger are found in different parts of western Texas, and are more of less destructible to all kinds of stock. They are frequently killed by our citizens and huntsmen.

COON, OPOSSUM, ETC.—The coon, opossum, jack and common rabbit are as plentiful in western Texas and probably more easily secured than in almost any other country.

SQUIRRELS—Of the squirrel I have seen but two kinds: one, a ground squirrel much like the northern chipmunk, the other in size and color between the common red and grey squirrel of the north, which last are very plentiful in western Texas.

MINK, OTTER, ETC.—Mink, otter, beaver, muskrat and weasel I have never seen here, but am told that they are found in the upper country.

FOX—The fox is in the upper country and of different colors—silver gray and red, I am told.

POLECAT—The polecat is here all over any of our prairies; as I have heard it asserted the fur of these ill-smelling animals is found to be very valuable, a person could pay himself to travel our prairies and kill them for their fur.

How many of the animals that I have mentioned, and what other animals in western Texas, are valuable for their fur, I do not know, but I should suppose that the huntsman and trapper might do well here in pursuit of animals for their skins and fur.

Game Birds

Hunted in western Texas are the turkey, sandhill crane, prairie chicken, quail, plover, snipe, pigeon, swan, goose, duck, etc. Our fancy and songbirds are the mockingbird, redbird, whippoorwill, bird of paradise, turtle dove, etc.

Wild Turkey—The wild turkey is excellent for the table, and there are any quantity of them in western Texas. I have often been with crowds of stock boys when coming up on droves or scattering turkeys in open prairies. Here and there would go a man or boy for his turkey, but not as hard as he could run at first. If he understands the work, he will try and keep the bird on foot as far as possible, for the purpose of getting it wearied before it takes the wing. But when he flies, the rider increases his speed, and if the turkey's wings give out any considerable distance from the timber for which he is making, the horseman is almost sure to overtake and capture him. But if the turkey can gain the timber, he is safe, unless the rider has a revolver or gun, with which he can sometimes ride near to the tree where the turkey is and shoot him.

In the spring of the year, when the males gobble, there are thousands of turkeys shot in this country by the hunter camping out or going early in the morning into the bottoms or timbers where they are, and, on hearing them gobble, carefully approach and shoot them down from their place of roosting. They are sometimes made to gobble by imitating the hen, which is often done with the help of a peculiar bone taken from the turkey, and so fixed that, by placing it to the mouth, a very accurate imitation of the hen turkey is made. Many a strutting gobbler, in responding to the call of this bone, brings upon himself the unerring level of the hunter's rifle. Most people, with a little practice, can imitate the hen without the aid of the bone. Turkeys are here frequently hunted with dogs, trained for the purpose of flying them into trees, when the hunter shoots them down. They are found high and low wherever there is timber.

SANDHILL CRANE—The sandhill crane is a bird somewhat larger than a turkey, with much longer legs and neck, and no tail. They are of two colors—white and light gray—the former being a little the largest. During the winter and spring they are very much numerous upon our prairies. Where they go to hatch, I do not know. They are rather shy, although plenty of them are shot; and when properly cooked, they are fine for the table.

PRAIRIE CHICKEN—Prairie chickens are nowhere more numerous than here, and they are easily shot by anyone who knows how to handle a gun. I have repeatedly seen a gentleman and his wife riding in their buggy along the road or through the prairie, and every now and then stop their horse for the purpose of shooting a prairie chicken which their pointer had set, and which the gentleman would shoot while standing in his buggy. After picking up the bird and reloading the discharged barrel, the gentlemen would get in and drive along until soon the fixed position of the dog would bring them to a stand, and bang goes the charge for another chicken. A dozen of these birds are soon obtained in this way upon our prairies. Their nests are frequently found with a dozen or more eggs, which sometimes afford the stock boy a hearty meal.

The New England quail are plentiful here, and are a pretty mark for the hunter, and there is no more tempting dish than their plump little bodies make. A prettier sight there never was than a couple of quails with their dozen little ones, just off the nest, tripping upon a sandy level. I have never seen them driven into nets, as in Illinois. I believe they are a little more shy here than in the North. Like the prairie chicken, they are found all over western Texas.

PLOVER, SNIPE AND PIGEON—Plover are here in abundance but near the coast, principally, I believe. They are a plump bird, and pretty game. Snipes are everywhere, fat and fine, in western Texas. Wild pigeons are seldom seen near the coast, but are more or less plentiful in the upper country. I suppose the more general production of the cereal grains nearer the coast would bring them there.

SWAN—Swan, geese and ducks are here during the fall, winter and spring, in such numbers as were never seen in any country but this. It is not unusual in spring to see our bays and lakes for miles out, and along the beach as far as one can see, covered with millions of these water birds. The beautiful swan is more shy than geese or ducks, but a good marksman may kill thousands of them. Their downy skins are very valuable when well preserved, and why a good marksman could not make money here at the saving of these beautiful, downy skins, I do not know. Their skins are worth several dollars each. It is said that when coming upon a flock of swans on a windy day, when out upon our prairies, if they are suddenly rushed upon on horseback, and driven with the wind, they cannot fly up or rise from the ground, and they are easily shot in this way.

GEESE—Geese frequent, in flocks of hundreds, our short-grass prairies, and fatten upon tender grass and herbage. They can be seen by thousands at a time in different flocks upon our prairies. They light near dwellings, where the grass is short and sweet, and where they can nearly always be found for weeks together. They are a little shy, and know pretty well the distance to keep from an armed man. But anyone can shoot them, and a good marksman can take down as many as he wishes.

They are of two kinds, and are here called brant and geese. The brant are the dull blue color, and their noise is more like the noise of the domesticated goose of our country. The geese are of a brown color, with black about their wings, breast and head. They are not considered first rate eating, though when well cooked and with the right kind of dressing, they are very good.

It is not long since I saw a person who had a hat lantern constructed, with which to bewilder geese in the night. With this "torch hat," as he called it, he said he could go among them and kill them with a club. He said he could get them in ponds of water better than anywhere else—the reflection of the rays of light upon the water having a peculiar effect upon their eyes, I suppose, is the reason for this. Should they fly up before he could get near them, he said they would come all about the light, which would enable him to occasionally strike one down. I have no doubt that the hat-lantern, torch-light or whatever it might be called, would prove successful in approaching geese in very dark and foggy nights.

I myself have made fires in thick clusters of trees near to ponds of water in which there were geese, and after frightening them up they would fly about the motts, and frequently come so low that they would strike the trees and come flopping and tumbling down through the branches to the ground. Of course, when striking and tangling in the trees, the flopping of their wings could be heard, which would enable one to get nearly under their place of falling, and ready to pick them up or strike them with a stick. Sometimes, when striking the trees with their wings spread, they would lie where first lodged, upon the dense foliage, for a half hour or more, dazzled by the light of the fire through the branches. When tired of their position, they would commence flopping and down they would come. It takes a dark, misty night to do this, and in such a night, a number of persons could soon procure enough feathers to make a fine bed, and as many geese as they would wish.

I have several times, when cattle hunting, camped near ponds of water in which there were thousands of geese. During the night, wolves, coons or animals of some kind would, every little while, slyly rush in upon the geese and keep them flying and squalling, and often succeed in catching them.

Let the reader imagine himself, on a dark, misty and stormy night, out upon a prairie in a cluster of trees, in which there is a large fire. It being a wet time, the ponds all about these trees are full of water, in which there are thousands of geese and ducks and loud-voiced cranes are about. Now, suddenly, rush into these ponds upon your horse, among these geese and ducks and start them up. Soon they are flying in every direction and the air is alive with these squeaking birds. The strong winds hurl them here and there in their efforts to rise and contend with the furious gale. The bewildering blaze brings them to and fro about and above the cluster of trees. Now, being lose and allured by the glimmering of the fire, they come too low, strike the trees and fall an easy prey to those underneath, provided they are keeping up a sharp lookout and get ready for the fall. Upon such occasions, I have known sandhill cranes to join in the dark and noisy confusion, and come flopping down through the trees and run directly into the fire. This, no doubt, would be new to many Texans, but anyone who will take the trouble to try the thing, on a dark and stormy night in a suitable place, will find my statements correct.

DUCKS—Wild ducks are of many kinds and colors in western Texas, and their size is from that of a bantam hen up to that of the smaller brant. Many of them are not only beautiful in color but excellent for the table. When returning, in the fall of the year, from their periodical northern flight, they go every day to the mast-producing timbers and fill their crops with acorns. When the acorns, etc., are gone, they live upon tender herbage and grass. The little boys of the towns or country go out upon their gentle ponies, from whose backs they can safely fire their guns, and soon return with fine lots of ducks, geese, etc., etc., hanging to the horns of their saddles. For many months of the year they are very plentiful in and about all the fresh waters of the country. They are also along the coast in the bayous, salt water ponds and, as I have said, line the beach of the bays. I am told that they lay along the shores or shallow waters of our bays and live upon the little fish and what they can find in the muddy bottom.

SONGBIRDS—Of our songbirds I will mention only the mockingbird, which is indeed the most interesting of the feathered tribe in all the land. If the western Texan's dwelling is well surrounded with shade trees, he is not only blessed with the music of these birds by day, but can hear their wonderful mimicry the livelong night.

FANCY BIRDS—Of beauties I will mention the redbird, whose song is sweet but without many variations. They have become so tame, when about the dooryard, that they will come to the door of the house and pick up crumbs of bread within an arm's length of you. My little boy has had them so tame that they would come every day from the timber to the door of the house for their feed of crumbs.

Ugly Birds—Of our ugly birds I will mention the buzzard, which is valuable as a scavenger. If an animal dies upon our prairies, these birds are the first to discover it; and many a dead animal is found and a valuable hide saved by the stock grower, who observes these birds as they collect and fly over the dying or dead animal. It is said that these birds fly to tremendous heights every day of their lives. They have a singular habit of standing upon branches of trees, fences, etc., and holding out their ample wings, perhaps for an hour at a time. Does not this, and their high flights, indicate that they need considerable airing in order to be agreeable to their not very sweet-smelling selves?

The egg of the buzzard is much like that of the turkey; and although the color of the grown bird is dark or quite black, the young buzzard is white and a comical looking thing it is.

The Eagle—The king of birds, both the gray and bald eagle, are here; and in motts and lone trees out upon our broad prairies, their nests of sticks and coarse material are found. Here their shrill notes are heard in defiance of everything but man, and when he approaches to disturb their nest of young, they hover around as if to say, "Thou art my only dreaded foe."

THE CHRISTIAN MAN

Who doubts that this is the place to live? Although the luxuries and refinements of older countries are not so much to be found; although some of us who are here might be better than we are, and a better state of morals and more decency might exist in some places; yet, education, religion, society, morals, etc., are fast brightening up and are certain to do as much for western Texas as these great elevators of man can do for any part of the world.

And it is a good country for the good and Christian man. And why? Because there is plenty of room here, and the good man can be unmolested in the enjoyment of his religion and property, and his example would exert a good influence upon those around him. Because the good man's character would here, as in other places, stand transcendently out, he would be looked up to and respected for his virtues. There are plenty of men who, if not very good, have a reverence for the moral man, and would act uprightly and conduct themselves properly in his presence and within his knowledge, when otherwise they might not. I suppose no one would deny that a greater moral restraint would make a happier country of this, and that more truly moral men would benefit our communities, which should make it an inducement for the Christian to come, and a satisfaction for him to live here, since he could do it so profitably to himself and others.

The Rich Man

This is a good country for a rich man, because he can here invest his money with more certainty of a fine income, with less attention and feeling less anxiety for his safety, than he could in almost any other country. He can here purchase property, not with fear or apprehension of a depreciation, but with a certainty of its advance in value. He could here, if I may be allowed the expression, be a king among men and not a man among kings. Gentlemen, or moneyed men, are living in nearly all the towns of western Texas, who own large stocks of cattle, horses, etc., upon our prairies and who sometimes own a ranch, riding horses, etc., for the management of their stock, and again they pay by the head, or give a share of the increase for branding and the little attention their stock requires.

The Merchant & Planter

This is a good country for the merchant because the profits of trade and traffic are large, and the mercantile business, well managed here, is sure to make a person rich in a short time—at least, so it seems to me. I do not exactly understand the reasons why, but certain it is that merchants, as a general thing, do better in western Texas than I have ever known them, as a class, to do in any other country.

I have heard it remarked by merchants here that stock growers are their most ready cash customers. It is not unusual for the cotton crop of western Texas to fail and, consequently, cotton growers frequently run long accounts. But the stockman, always on hand with the money or a crop of beeves, mules or something that will readily bring the cash, keeps the merchant on his "taps" until the planter makes the crop, which enables him to settle his long-running account.

Cotton, for the last season or two, has done better and has raised the drooping hopes of many who had almost given up western Texas as a planting country. How much of the country west of the Colorado River bottoms will prove itself as reliable for cotton, I suppose is not yet determined. I have always considered western Texas as a great southern stock country, and I have yet to change my opinion. But if the seasons were generally different, I suppose there would be no better cotton lands than the river bottoms here. I am acquainted with a large slaveholder who, years ago, abandoned the raising of cotton in one of the best localities in western Texas, as too uncertain a business for him, as he said. He turned his force to the raising of mules, cattle and the extensive production of pork, bacon, etc. I have talked with other large and reliable planters of western Texas, who say they would not exchange their localities for any that they know of in the older cotton growing states, and probably not for any in the world.

It may seem a contradiction to say that the merchant of our country does remarkably well when the cotton grower, who should be his support, does not so well.

But the truth is, cotton growers are also, to some extent, stock growers, and of late better crops have been produced. It is to be hoped that the seasons of western Texas may continue favorable for the production of this king of crops, and call to the country planters to occupy all its good cotton lands. Should the seasons of western Texas continue favorable for cotton, planters could not do better than to come here and secure the rich bottom lands of the country.

A good share of our merchants, in addition to their stores, have stock ranches, and not infrequently exchange goods for stock. By employing a stockman to attend to their ranches or their stock, if they have no ranches, they carry on the two branches greatly to the advantage of each other. The merchant frequently concerns some sharp-eyed fellow with him and sends him off, with a little money, to Mexico for a drove of horses or mules. He sometimes sends on, in charge of an experienced, trusty fellow, a drove of horses, mules or beeves, to the north. In purchasing such a drove, he frequently collects outstanding accounts or exchanges the goods for animals, and often making up a drove, by hook and crook and getting it well on the way, he goes on by water and rail in time to meet his drove, at some designated place. Turning the drove into cash or good short paper, to the best advantage he can get at Chicago or some other point, he goes on to New York and pays up, makes a fresh purchase of goods for his store, and comes home.

Merchants also purchase and exchange goods for various products of the country, such as hides and peltry, wool, pecans, moss, etc. They also engage, more or less, in shipping beef and veal cattle, mules, horses, etc., to New Orleans, Mobile, Cuba and other places.

Previous to the last two seasons, I noticed that it was not an unusual thing for a merchant to get a mortgage upon a plantation and negroes in western Texas. But I have never yet known a stockman under the necessity of mortgaging his property in order to make ends meet. This is not because the stock grower does not go in debt, but because when, from any imprudence or from going in debt for cattle, lands or to his merchant, finds himself troubled to meet his monthly payments, he can do as I have repeatedly told him. But the planter, when he gets pinched, cannot sell his cotton and sugar until it is in the bale or hogshead, unless he pledges his crop before it is grown and then, from drought or some other cause, it might fail, leaving him in the merchant's debt. His expenses are heavy and his debts increasing, which too often results in a mortgage upon his plantation and negroes.

There are other reasons for the ill success of the planter in western Texas. In coming to the country with his force and family, he incurs a large expense, and in the purchase and improvement of new lands, building, etc., another large outlay follows. His account, upon arrival here, commences with his merchant, depending, as he generally does, upon his first year's crop to meet his engagements, which crop is to be raised in a new climate where the seasons are different, and upon new land, the like of which he is unaccustomed to tilling and the correct management of which he does not exactly understand. Having relied too much upon his ability as a cotton grower in the old state from whence he came, and consequently being a little extravagant and regardless of expense, he often finds himself in debt, with perhaps but a fourth of a crop of cotton, which barely pays the expense of raising. Now comes the landholder for the balance due upon his negroes and other property, which staggers the planter and leaves him but a short and unsuccessful career.

Now, although I am out of my place, I will venture to give the planter who comes to western Texas a little advice, that may not be to his disadvantage. When he arrives, let him be moderate in the purchase of lands, which should be of the rich river bottoms, with perhaps a strip of prairie. Let him invest, say one-half or a portion of his means in cattle and horses, and of course he would get a stock of hogs, and if in the right kind of range, he might get a flock of sheep. He should be careful how he meddles with sheep unless he has a great taste and inclination for the business.

Now we will suppose, that with this kind of start, his first and possibly second crop of cotton fails. How is the planter to get along? Is he to be broken up? If he has been prudent and economical, it seems to be, not. Although his cattle, horses and sheep might not at first afford him much of an income as would abundant crops of cotton, in case of failure of his crop, his stock would be his salvation. And again, should the country or his particular locality eventually prove too uncertain for the growth of cotton, could he not plant his rich bottom lands in mesquite and bermuda, and still be "king" in the way of stock grower?

I have known several planters to come here and go in debt, as I have stated, and from the failure of crops become largely involved, when, if they had invested a portion of their means, or parted with a share of their force, and put the proceeds thereof in stock, they would have undoubtedly got along until they could have been favored with a good season for cotton, the large proceeds of which would have swept their debt away.

I do not say this for those planters who are here and know ten times as much about the thing as myself; but for those who may come to the country in circumstances that could be easily embarrassed in this peculiar climate.

To the planter who has started as I have suggested, I say as I have said to the stock grower: If he finds himself cramped to meet his payments, let him turn to his stock and dispose, if need be, of all the male animals or beef cattle, young and old, from his stock, to the best advantage he can; part with his yearling mules, wether sheep and lambs, wool or anything that will not injure the increasing capacity of his stock. At the same time, let him watch old and successful planters, and make all necessary inquiries in regard to the proper management of the cotton crop in western Texas. Let him work early and late, breaking up and preparing his lands for a good crop, which by perseverance, will surely, sooner or later, bless him and enable him to square his yards.

Would my remarks discourage the emigration of cotton growers to western Texas?

Let us see how this really is. In the year 1860 we will bring Settler A from Alabama to western Texas, with a force of negroes, and settle him upon a certain river where he shall improve a place with the expectation of getting a crop from the first planting. He contracts debts with the idea of paying them with is first crop of cotton. Feeling sure of a crop, he has perhaps made some debts that he could possibly have avoided. His crop, from some cause, fails. He gets the payment, now due upon his plantation, put over to next year and his merchant is also put off until the next crop. The next year he gets no crop, or but a slim one, and he begins to feel the weight of his debts, which are presented for payment, and he has nothing with which to meet them except his negroes and the plantation with its appurtenances. The result is bad. Settler A and his family, overseer and all, write home to their friends in Alabama not to come to western Texas, and tell them their troubles, the whys and wherefores. Their correspondents, and all who hear from them, spread the gloomy tidings from western Texas, and people stay at home.

But let us, in the same year (1860), bring Settler B to western Texas, and settle him upon the same river. He also goes into debt and improves a plantation. But he does not buy as much land as did Settler A, instead of which he buys a stock of cattle, mares or sheep, and possibly brings back with him a jack and a few bucks, for the purpose of raising mules and improving the flock of sheep he may buy. His first and second crops of cotton turn out as did those of Settler A. He now depends upon his stock, which he sells off, as I have directed, as close as it will do without injuring its increasing capacity. Settler B makes another effort for a crop, which results most favorably. The consequence is that he and those about him write home to their friends and say, "Although our cotton did not do so well for a year or two, our stock kept us along, and this year we have made a tremendous crop, which has set us way ahead. We have learned the country and a glorious country it is. Come to western Texas, my friends, and do as we have done. You will never be sorry." The result is that friends of Settler B and family circulate the glad tidings from western Texas, and more or less of the recipients of these good accounts make their way to the country.

Is this discouraging the emigration of planters? Or is it putting a flea in the ear of the planter who might come to western Texas and run through his property, for the want of just a hint as this, and thereby retard the settlement of our cotton lands?

And again, you southern boys, please read my book, which may seem to have been written more for the Yankees, but only for the reason that I happen to know more of them than of you, and know them to be the people to make a great wool growing country of this, and develop its resources and importance in other respects; but in many of my remarks I have used the term "northerner" to designate people north of Texas.

I would not apply my remarks in regard to the failure of the cotton crop to eastern Texas, nor even central Texas for, if I am not mistaken, the Colorado bottoms are equal to any lands in the state for the growth of cotton, and the seasons along this river, I believe, are uniformly good for cotton.

Now to return to the merchant, whom I have purposely mixed in with the planter and stock grower, as he is so peculiarly interested in the whole country...I said that the profits of trade in this country are large.

This may seem in conflict with much that I have before said, unless it should appear that the profits of all kinds of business are correspondingly large, which is generally the case. It is a fact that the merchant's profits upon his goods in western Texas, compared with the northern merchant's profits, are as a half dime is to a penny or thereabouts; and this may be said of mechanical business and trade generally. What is bought in the north at wholesale for a penny and retailed at two pennies, is here retailed for half a dime at least. Were such a ferry as the New York and Brooklyn established anywhere between two cities in the south, however large they might be, the cost of a ride from one to the other would probably be five cents, instead of a penny as in the north. The blacksmith, for shoeing a horse in western Texas, will have more than double the amount the smith gets for the same job in the north.

This is a true scale of prices upon which business is done in this country, and why should not merchants and tradesmen do well here? If a trinket of a penny's value, in a mercantile or shop establishment of the north, would here sell for half a dime, and often a dime, and if the foregoing is illustrative of business in general, what say you, merchants and tradesmen of all kinds, to this being a good country for you? It may justly be said that it would cost more to run a ferryboat, to get articles of trade and the blacksmith's iron, etc., in the south than in the north, but this difference of cost of raw material in the two sections is nothing to compare with the different in profits.

The penny or cent is rarely seen here. I do not mention this as a credit to the country, nor as a discredit to those states or countries where they are used. It is only to show the profits of trade in western Texas that I speak of this coin. I believe that a coin of this small value is a good indication. It is a proof of enterprise and industry. We will suppose that the penny coin of New York City might be done away with, and that nothing less than the five-cent piece were there used, what would be the profits of the Brooklyn and New York City ferries when getting five cents for what they now have but one cent, and with which they make money?

The great necessity of the penny or small coin in the north is a striking proof of economy, and, more than all, of ingenuity. Were it not for the invention of machinery, and the ingenuity of the machinist and shipwright, how could such crowds of people go in the convenient and quick way that they do, for a single penny, from New York to Brooklyn?

I would not make people small, but if ever the time is when the penny will be indispensable in the south, it will be when there is more productive skill and ingenuity among her people, and when the profits upon articles of trade are not as large as they are now, or when more business is done to realize the same profits.

SPECULATION & CATTLE DRIVING

Western Texas is a good country for the speculator, so far as articles of speculation are produced. Because, by buying here, the speculator has not only the chances of a rise in the price of his property, but the advantages of buying in a country where things are easily and cheaply produced, and such things, too, as can be cheaply exported to those countries where they are more laboriously and expensively produced. For instance, a beef here can be grown to a weight of six hundred pounds at a cost of two dollars and a half, and driven for the same amount to a country where it costs, perhaps twenty-five dollars to raise a beef of the same weight; and from that, with the twenty-five dollar beef, on the same market at the same expense. This, no doubt (although the twenty-five dollar beef is worth the same money) would result in a big advantage to the purchaser of Texas beef over the purchaser of twenty-five dollar beef—provided, the Texas beef is driven by the right kind of experienced drivers. And when the cattle of Texas are improved, there will be a still greater advantage in favor of the purchaser of Texas beef cattle, if speculation or driving north be the object of the purchaser.

In fact, if those northern men who now come here and take beeves to Chicago and northern markets would manage differently, they would make much more money than they do now. If, in the place of taking wild, rough, light-quartered animals, they would take choice cattle only, and employ experienced managers and drivers, they would, at anything like the present prices in Texas, always be very sure to make money, and with good luck, their profits would be large. And by always demanding choice cattle, the stock growers would see the necessity of improving their cattle, which improvement would make a demand for blooded stock, which the drover might take on to Texas with him when he goes for a drove of beeves. Should a northern drover think of doing this, he should go on to Texas with his fine stock in the fall, and sell out during the winter, then get ready to start in early spring with his Texas beeves for the north.

A drove of about seven hundred beeves was driven from middle Texas to Chicago, about two years since, which it is said cleared about ten thousand dollars. This drove was no doubt well managed, but there was a large number of inferior cattle in it, and had they happened to have gone into a bad market, these inferior ones might have caused a loss of money upon the drove. Whereas, if the whole drove had been choice cattle of our country, they might have gone into the same supposed bad market and still have made money.

To my knowledge, there have been many men here, within the last six years, who have bought large droves of Texas beeves for northern markets. Some of these men have not done well, simply because they did not know how to buy, select and drive Texas beef cattle. When a northern man comes here for a drove of beef cattle, let him come in time to look around and see where he can buy to advantage. He need not make any noise or let people know what he is about, until he finds a large prairie or extent of country where the beeves, for the year or two previous, have not been much culled or sold out. When found, he should secure such cattle, unless they are known to be generally wild and bad to manage.

There are certain prairies in Texas where northern men should be very careful about buying; and wherever, or of whoever, he may buy, his contract should call for all of his or their beef cattle; that is to say, all such beef cattle as he, the purchaser or his assigns, shall consider good and merchantable. Then, in his selection, he should not take a light, rough, washy or ugly beef, nor those that are at all staggy. But he should select those that are uniformly smooth, with as clean necks and horns as possible; and if the judge or selector is capable, those that would be most likely to drive and feed kindly, and above all, those that are well-quartered, or put up in a way to take flesh with fat and be sure to weigh well.

Of course, the drover or manager would have his men ready, and most of them with him, and his contract should call for the delivery of the cattle across one large river or stream, at least, at some point where the balance of his men should be in readiness to help count and receive them, and where the wagon, camping material, extra horses and the whole complement of drovers, cook and all, should be in readiness to move the cattle on, or hold and herd them here at headquarters, until perhaps the drove is made up with cattle coming from other points.

At this stage of the enterprise, if the manager and drivers understand their business, they will use the utmost caution, and give all possible care and attention, to avoid a fright or stampede. For a week or fortnight, if the men study their own good, they will use the greatest caution and do all they can to help the sleepless, wearied and care-worn manager prevent a fright or stampede, and watch closely that none may get away—after which time, if all has gone right, there is not much danger. Yet, unless a man is all attention and is proud to do his duty, he should not be allowed a place among the faithful, hardy and trusty drivers of a Texas drove of beeves.

Whoever starts a drove of our cattle north should, by all means, employ an experienced Texas manager, and consult him in regard to the kind of drivers to employ and possibly in regard to the purchase of the cattle. The place for receiving a drove of Texas beeves, or for holding them or a share of them until the drove may be completed or ready to start, should be where wagons or men on foot do not frequent, and, if possible, where nothing might occur to frighten them. The writer has driven these Texas beeves for several years, and for the last two years, in driving many droves, he had but one stampede to speak of. This he would have avoided had he not, necessarily, been away from the drove.

A northern or western drover should always receive and start his cattle from Texas during moonlit nights, if he possibly can, and as early in the spring as the grass will do, when the weather is almost sure to be fine. By moonlight the drivers can see what they are about when around the herd, and by the time they have to do without the light of the moon, the cattle should be well broken in, so as to drive and manage without difficulty. A drover should be as sure as possible to secure the right kind of men, and besides paying them the wages agreed upon, he should treat them as liberally as he consistently could, and then he should expect them to do their duty. If he is inclined to be penurious, let it be with someone besides his drivers; for a lot of drivers will do any amount of mischief to the employer they do not like, and vice versa, they will watch over and almost die with the cattle of any drover with whom they are pleased.

I mention these facts simply because I have known northern men to come here and not do as well, by several thousand dollars, as they might have done in the purchase of their droves, and then, to save money, would employ inexperienced hands and undertake to carry the thing through on the cheap or stingy principle, and consequently, get into trouble and make nothing.

The true way is to purchase the cattle as shrewdly, closely and judiciously as possible, and then compensate the drivers liberally. Save money out of anyone but the drivers, and if you are the man for the business, you are sure to make money in the end, and that largely, too. By treating the drivers liberally and well, I do not mean that you must give them their own way or let them impose on you. Use them right and they will suffer you, as it were, to see that they do their duty.

The purchaser should remember that he is not going to market every week or month, but that he has a long way to drive, and that it costs no more to drive a choice Texas animal than it does an inferior one, and the chances are ten to one in favor of the choice animal. For five long years I have bought and driven Texas beef cattle, and I have made a good deal of money for those whom I worked; and I know that the right kind of men can make money, with as much certainty as there is in human affairs, by driving Texas beef cattle to northern markets, but they must be "up and dressed" for the business. I have recently heard of one of five or six partners, who were each more or less interested in three droves of beeves that were driven from western Texas to Chicago last spring; and the partner of whom I speak wrote home from Chicago that his share of the profits on the two thousand cattle they drove would be over nine thousand dollars. I believe there were at least two other shares in the company whose interest was as large or larger than his. In this large lot of cattle were the eight hundred head of which I spoke, near the beginning of this work, as being the sale of stock grown at sixteen dollars per head.

Cattle Shipping

The shipping of beef and veal from western Texas to New Orleans, Mobile and Cuba is an extensive and, to those who are calculated for it, profitable business. The writer is largely experienced in this trade, to which he owes principally his knowledge of western Texas. It is the buying, driving and shipping of beef and veal cattle from different parts of Texas to New Orleans, for a wealthy company that has, more than any one thing, given the writer such an extensive acquaintance with the pursuits and inviting condition of the country to foreigners.

There is room here for competition in this business, and it is to be a large and growing trade in western Texas for, I might say, all time to come. The writer could enlarge to almost any extent upon this subject, but as no great good would arise from such enlargement here, he will simply say that not a few have made and are making money at this business, and that Yankees and Germans are among those who are doing as well as any in the trade. I will further say that, without any experience in Texas, I commenced this business at $60 per month, and soon got to $75, and then $100 per month, which situation I gave up to write this book. Had I continued, I probably could have commanded any reasonable price for my services. I mention this merely to show what a verdant man or plowboy from the north can do with the wild stock of Texas—i.e., if he is of the right stripe, and not afraid to work.

TEXAS

Beef Packing

Packing beef has been often attempted without success, but recently it has been successfully carried on, to some extent, in western Texas. It is a well known fact that if a process were generally understood whereby beef could be packed in Texas so as to keep and pass inspection in our large markets, it would soon be an extensive and, no doubt, profitable business.

There has been, to the writer's knowledge, within the last three years, several hundred barrels of beef packed in western Texas for the New York market, which bore inspection, and did very well for the packer, although he labored under great disadvantages in the operation of packing. The beef packing business in western Texas deserves the attention of enterprising men. That it pays to drive beef from Texas to Chicago, thousands of miles, and pack it, has been well demonstrated. Why should it not pay to pack it in Texas, if the thing could be extensively done, and ship it instead of driving so far to the great markets of the country?

The writer has recently learned that a gentleman has just come on from the north to Texas to invest a large amount of money in this business. He knows a process of packing that will no doubt insure success in the business. We have several months of cool weather in winter that certainly ought to admit of profitable beef packing.

Speculation in Horses, Mules, Etc.

Horses and mules have been, and are yet to a great extent, driven north, east and west from Texas. They are so cheaply produced here, and the work of improving them will be so rapidly carried on, that soon the speculator will be able to get fine horses and mules in western Texas for one-half the amount it would cost to raise such animals in some of the northern states.

These animals are more easily managed and driven than cattle. Our horse drovers, after getting the animals accustomed to each other and well broke in, frequently, when driving through our open countries, turn them loose without a herder when night overtakes them, and generally find them all together, not far off, in the morning. But great care and watchfulness should be exercised until the animals are reconciled to each other, and are inclined to stay together or follow the bell-man of the drove. Sometimes mules will unavoidably stampede from the scent of wolves or other animals, and various other causes. But an experienced manager is at home with them, and seldom meets with any loss. Such a manager should always be employed with a drove of our Texas horses or mules.

The immense amount of this kind of stock in Texas and Mexico, its rapid increase and the cheapness of its production, should certainly call the attention of speculators this way. I know of a Missourian who has, for the last eight years, taken a lot of yearling mules from Texas to his farm in Missouri, and after feeding them on corn for a season, has taken them among the planters of the south and sold them, generally at large profits. He told me that he raised a hundred acres of corn annually, and that he disposed of it in this way—some seasons at several dollars per bushel. The bringing of those cheap Texas jennies to an abundant corn country, and raising half-breed jacks on corn for the Texas market, would not be a bad business. Fine jacks and stallions can be exchanged for Texas mules and horses at a profitable rate, provided the thing is properly managed. It would pay well for a speculator or stock grower to buy up old, foundered and used up mares, anywhere along the Mississippi states, where they could be bought cheap and driven to Texas, with a little expense, and kept there for the purpose of raising mules, or sold to the stock grower in exchange for mules, beeves or something of the kind. It would pay the speculator, who is watching for rare chances, to peruse well this book.

Land Speculation

From the appearance of things, the land speculator could nowhere do better at present and for many years to come, than in Texas; but this business is too much on the "shark" order to admit of much encouragement from a friend of the poor man. I think I could write many interesting pages upon this subject, but it would conflict too much with the spirit of my work.

Wool & Sheep Speculation

In the opinion of the writer, there will be in twenty years not less than ten million sheep in Texas, and possibly twice that number. In the meantime, would not the speculator be justified in looking that way for wool, mutton, etc.? There are those who could probably come here at this time and do well at the purchase of the common qualities of wool in small quantities. Those raising fine wool, I believe, generally know pretty well where to send it to get good prices or realize the most out of it. But many of those who have small lots of coarse-wool sheep frequently sell their wool by the fleece, at much less than value.

I will close this subject by suggesting that the wool speculator read this work through, and perhaps he will be better able to judge of the good chances and wide field for his operations in this country than myself. Certainly it would seem that the wool speculator might do well in a country where this article is so cheaply and easily produced. I will say that the speculator can hardly go amiss by taking sheep from the northern and western states, across the country, to northern and western Texas, provided he understands the business of buying, driving and managing sheep, about which I will here drop a few remarks:

All who are acquainted with the sheep business know the danger of overdriving them. When starting a flock for a long journey, the greatest care should be used not to overdrive them for the first few days. If you do, the very next day, the chances are that some of your flock will begin to lag and go behind, and are to be carried in the wagon and eventually sold for a trifle, given away or die. In regard to age, condition, soundness of feet, etc., any person of the least experience should know. Sheep in a four-rod road will nibble their way along as fast as they should go without being driven. In fact, there should always be a man ahead to keep them back when they are inclined to go too fast. When going through forests, marshes or places where there may be poisonous vegetation, sheep should be well watched, and if possible, they should be well filled with grass and not hungry when starting into such places, and they should be put through as soon as it will do. There should always be an antidote for poison in the provision wagon or near at hand.

When driving a large flock in an open country, in warm weather, they should be strung out and not suffered to huddle together too much, which heats them and often results in many wind-broken sheep, and is injurious to them in other ways. They should not be suffered to gorge themselves or drink excessively and then driven fast. When driving through an open country, the flock should always be guarded in the night by one man at least, and the shepherds' dogs should be in their places. If droves of sheep are not closely watched in our prairie countries, they are liable, in the night particularly, to divide up and stroll off in gangs. Too often, wolves get among them before they are found, and frequently it is impossible to trail or find them without troublesome and expensive search.

Lastly, but my no means leastly, when coming to deep creeks or rivers, the greatest caution should be used that your sheep are not drowned or injured by swimming. A healthy, strong sheep will safely swim a river of good width, provided it can have a good landing place and the current of the river is not too strong. A flock of sheep should never be put into a rapid stream, unless there is a broad, easy landing upon the opposite side, and then a few should be tried first to see how they make the landing. There should always be a person at the coming-out place to squeeze the water from the wool of any sheep that is too exhausted to get out without assistance. A large flock of sheep should be divided and swam in parts, or else it should be strung out and but a few allowed to go in abreast. And there should always be two men, at least, at the starting-in point, and should anything go wrong in the water, or at the landing, that would be likely to huddle them together or interrupt their swimming in the least, the flock should be instantly stopped and driven back. Great care should be taken that the balance of the flock does not take a run and get the advantage of you in their determination to follow those that, perhaps, are yet in the water and not out of danger.

A safe and excellent way to swim sheep is to have the flock, or a portion of it, upon the water's edge, where there should be a skiff, with two men in it—one to sit aft and row the boat, without throwing out the oar in a way to frighten the sheep; the other to sit aft of him, if possible, and to draw with a rope, one or two strong sheep from the flock gradually into the water, as others are inclined to follow, calling them in a low voice with the sound to which they are accustomed. The moment they begin to follow, start the boat and head it so as to strike the opposite landing pretty well up. Without some awkwardness, this way of swimming sheep cannot but succeed. In fact, it is a good way to swim horses and cattle, also, if you have an ox or animal that will lead well into the water.

I could say much more upon this subject, as my experience in the business has been quite extensive, but I have said enough to caution any inexperienced or careless person who may start out for Texas with a drove of sheep. I do not think that in driving from Illinois to Texas, through Missouri and the Indian Territory, there would be much swimming to do, unless it should be an unusually wet season. Before risking much by putting sheep into river, I should pay for crossing them on ferries or in some safe way, particularly if the flock is not generally in good condition and strong. In conclusion, I will say, what I have in substance repeated, that a drove of sheep, or animals of any kind, should always have a good, experienced and prudent manager, when driven such a journey as the one in question. Such a trip would naturally be eventful and emergencies might arise. Consequently, whether going from the north to Texas or from Texas to the north, with stock of any kind, a trusty and experienced manager should be employed.

I once knew a Pennsylvania drover who came to Texas and purchased a large drove, consisting of six hundred beeves, for the Philadelphia market. The stock growers all got the start of him in the sale and delivery of their cattle, and then, to cap his bad management, he hired shop boys, town loafers and inexperienced men to drive this large drove of beeves, simply because he could hire them cheap. The consequence was that his drove wandered off several times, and many of his beeves were never found. A lot of green hands, without the right kind of Texas stock-driving manager, will frighten a drove of Texas cattle, horses or mules, in a way to bring disaster upon the drover.

Had this Pennsylvanian employed a number-one Texas beef buyer, driver and manager, and had he consulted and advised with him in regard to the selection, purchase and delivery of the cattle, and in regard to the employment of drivers, etc., his speculation would have proved a paying and pleasant one; whereas, the result of his bad management, of course, brought down his curses upon Texas and everything in it.

Mechanics

Mechanics of various kinds, will find Texas, and particularly western Texas, an excellent field of operation. The country is prosperous, and building is going on very extensively. Besides dwellings, stores, warehouses, wharves and the like, there are cotton gins, gristmills, sawmills, various factories, bridges, etc., being constructed. Manufacturing of various kinds is quite extensively carried on, and its material extension in many branches is confidently anticipated. Why should Texas send her vast amount of hides abroad to be tanned, while she is so well supplied with the finest of tanning bark? There are a thousand and one suggestions of this kind that I might make in regard to this country, were it not that I have other matters under consideration with which to conclude this work.

I will say, in conclusion to mechanics, that the country is fast filling up. Emigration is continually pouring in, and an increasing demand for your capital and services is to be the result. And furthermore, that I know plenty of mechanics and mechanical laborers who have invested their profits and savings of their business and labor in stock of some kind and, giving it in charge of stock growers to brand and look after, have become important and wealthy men.

TEXAS

The Poor Man

Western Texas, of all others, is the country for the poor man, and also for the man who would retrench, or who is unable to keep pace with the prodigality of the times in the country where it may be his fortune to live. I have somewhere spoken of a cluster of trees as the place of my writing. Many of these pages were written in the open air, beneath this group of live oaks, in the fall and winter season. During winter, much of the time, a person may sit comfortably in the shade of a tree for the purpose of writing or anything else. Of course we have cold snaps or northers of short duration, and short spells of cool, chilly weather, but no snow in western Texas. Consequently it is a country where the poor man can get along without expending all he has for the purpose of building a house to make his family comfortable.

He needs not a dime to build, in our timbered sections, a house in which his family could be comfortable and respected, and where he could live until his business and prospects would justify the erection of a better and more expensive one. Aside from the very few and cheap housekeeping material that would be required to get along for the time, he could put all he might have into stock of some kind, and then buy a home to the best advantage. If it is not more than five acres to begin with, it will do until the increase of his stock, and what he can pick up in other ways, enables him to buy more. Small stock growers resort to various means of money-making, outside of the stock business, the object of which very often is not to draw from the increase or proceeds of the stock for the support of the family, or for any purpose, except some solid investment that will add to their wealth. Almost any landholder will suffer a poor man to settle upon lands suitable for stockraising purposes, with the understanding that the settler shall buy as soon as he is able, or at some stated time.

I will say that in western Texas, the titles of lands have been much in dispute, and are somewhat so now. It therefore behooves a person to consider the title he is to get before paying his money for land. That is often an advantage to the settler, from the fact that the contending parties will allow him to improve a place, with an understanding that he shall pay to the party whom the law shall decide is the rightful owner and from whom a good title can be perfected, which often gives the settler ample time to get ready to pay for his place by the time the successful party, in law, is qualified to convey to him an indisputable title.

There is, however, an endless amount of good lands for sale in western Texas, the titles of which are perfectly clear and undisputed. In fact, the courts of the country, within the past few years, have settle questions of law and firmly fixed the titles to a vast amount of heretofore disputed territory of Texas.

Now it is to the poor man, who arrives in western Texas with but little or nothing, except his head and his hands to get along with, that I would address myself for a while. I will hereafter mention some of the different kinds of labor for which there is a demand in western Texas. But besides these, there are many things to which the poor man, particularly the apt and hardy northern man, can turn his attention with but little or no money to start.

Where I was raised, I know of men who were in the habit of going into the deep forests of the country to hunt deer for their peltry and venison during the whole winter season. They would travel upon their snowshoes from one month's end to another, separated from all human habitations by scores of miles, and were well satisfied if they could get the hides and saddles of one or two deer per day. These they would lug upon their backs for miles into camp, and perform more hard work in a tedious, snowy country to secure ten pelts with their two quarters each, than it would require to secure a hundred in this delightful country. If a person will use the right means in this country, he can make more money at this business than he could in any country I have ever seen, and I have seen many parts where deer were considered plentiful. And the beauty of the thing is that here it is not cold, tedious and disagreeable.

The hunter could here settle in some secluded place, where he could go on foot and do well, but with his gentle Mexican pony, trained for the purpose, do much better; and where he could go with a slide or stone-boat, drawn by a gentle yoke of oxen, and secure ten or twelve deer per day, the hides of which he could tan and get fifteen dollars per dozen for in a northern market, if well tanned, and the process of tanning is very easily learned. Deer skins are dried and packed away untanned, for the merchant, to a great extent, in this country. The carcass of deer, when so plentifully secured, unless the huntsman were living near some town, would not sell, but his family could be well and constantly supplied, and a drove of hogs could be fed or kept about home with this surplus of wild meat.

These remarks, I am well aware, would seem to conflict with remarks that I have made and may yet make. But there are, in all countries, men who will seclude themselves for the purpose of hunting, and why should not they be in Texas as well as anywhere else?

Would the hunter doubt his ability to make money in a country where there are millions of deer, and where, upon many of our prairies, there can be seen droves of hundreds every day, and where, on horseback, by looking in different directions, a person can sometimes see five hundred or a thousand at a time? How often I have thought, when traveling over this country, that hundreds and thousands of huntsmen, in different countries, would be delighted to know what they could do here, and would come if they knew, and convert these wild animals more to the use of mankind. It certainly seems a pity that they should live in such numbers upon our prairies, without the least expense to a mortal man, and left so uselessly to die with old age, or fall prey to wolves, panthers, tigers and other useless animals, when they might so easily be secured and made to answer a better end.

Huntsman, why not come here and send a bullet to the hearts of these old bucks, and make soft and beautiful leather of their skins, and send it to your cold countrymen to clothe their hands, in the shape of mittens and gloves; with which to make soft shoes for the old and feeble, and whiplashes for the drover, or notelayers for musical instruments, and a thousand other purposes? Why not come here and secure millions of these geese and ducks, and send their feathers to your cold countrymen, to make warm nests for their shivering maidens, in which they may sleep comfortably, when the cold frosty winds whistle around their chambers? Come and send home the snow-white downy skins of the swans, to comfort the cheery-cheeked daughters of icy lands. Come and decoy our animals of fur, and make use of this great waste of everything.

The different towns of the whole country require more or less hay and wood, and the grass upon the prairies is free to anyone who has a mind to cut and make hay of it—and who knows how to swing the scythe and manage the mowing machine? Many a little load of prairie hay have I seen sold for ten and fifteen dollars; and many a ton of northern hay have I known to sell here, in our seaport towns, for forty dollars; while the interior of the country is covered with mesquite grass, which makes the best of hay, and all around these coast towns there is heavy sedge grass that will do very well and is now taking the place of northern hay. Mowing machines are being brought on, and quite extensively used. Farmers and stock growers save more or less hay for the working stock that they wish to keep up and handy by. They give employment to these machines. Although a pasture of mesquite or bermuda grass, near at hand, would obviate this necessity, men are slow to cultivate grasses in a country that has always yielded abundantly and does now, except immediately around extensive ranches and farms that have been occupied for a long time, where the grass is eaten out.

There is an immense amount of freighting done in the country by ox, mule and horse teams. Many a rugged man gets his start at the stock business by freighting goods into the interior. It is not unusual here to see as many as twenty heavily loaded wagons, in a string, drawn by from four to six yoke of oxen each. Mexican horse and mule teams are considerably used, but more by the Germans than other people. The government freight business is very large in western Texas. Heretofore the government has done its own freighting from the coast into the interior, but recently this business, in the lower country, is contracted out, and the contractors employ Mexicans or re-let their contracts, in parts, to those who own a large number of Mexican carts and oxen, which are driven by Mexicans. It is not unusual to see a string of thirty or forty Mexican carts, all loaded for the interior, with government stores and supplies.

The oxen are yoked by the horns. These Mexican ox yokes are nearly a straight stick, grooved a little to fit the head, just behind the horns, and snugly lashed to them with rawhide straps. It seems rather cruel to men of the older states to see these animals work in this way, but the weight they pull, the distance they travel and the amount they will endure in this peculiar yoke is astonishing to all. It is nevertheless a wicked way to use oxen, as holding the head in one position for many hours at a time is evidently very painful, and the appearance of the ox is always that of great suffering. Mexicans are becoming a little Americanized in this respect, and are adopting the American way of working oxen.

It takes some little capital to get at the freighting business, but a man of good character can generally get together a few oxen and a wagon on time, and get at the business with but very little money. There is always job work of many kinds to be done in the towns of the country, such as street-making and repairing, garden-making, building fences and getting material out of the timbers for fencing, breaking up prairie lands, furnishing the towns with vegetables, butter and milk, poultry and eggs, game, oysters, fish, etc.

The gathering of moss, with which the forests of portions of western Texas are loaded, and which is used to vast extent as a substitute for hair by the upholsterer and manufacturers of innumerable useful articles of furniture, etc., can be done by anyone. This moss is prepared for sale and use by rotting and washing of the covering of its fiber, which resembles in appearance and answers to a great extent the same purpose as the hair from the tails and manes of horses. The rotting is done by burying it in the ground for a length of time. It is then taken up and washed, picked, pulled and cleaned, until fit for use. Thousands of people buy moss, in the shape of mattresses, lounges, etc., who do not know what they are buying, as it answers the purpose very well and is often sold for hair.

The gathering of pecans in western Texas, during the fall and winter seasons, often pays the poor man handsomely. Men with their families, children and all, turn out and gather hundreds of bushels, for which they get from two to four dollars per bushel. Many a poor man and family have gotten their start in this way in Texas, and become wealthy stock growers and farmers.

I have before stated that the forests of this country are tangled with grapevines which yield abundantly juicy fruit. There are those here who are turning these grapes into money, in the way of wine and brandies, any amount of which can be made, and the grapes free to all.

There is to be any amount of hedging done in western Texas. Unsuccessful attempts have been made at hedging, or rather, those who have contracted for the growing of hedges have abandoned their work and left it for others to complete. The reason of this, I suppose, has principally been the want of a knowledge of our climate and soil, which would undoubtedly have enabled the contractors to complete their contracts. Those who have continued the work, I hear, have made out first rate for themselves and have succeeded to the satisfaction of the planters and stock growers for whom the hedging is done. Of course, the Osage orange is principally used, and is no doubt the best plant in America for this purpose.

There are landholders, stock growers and planters in western Texas who would, no doubt, rather build their own hedges than contract for the rearing of them, provided they could save anything by doing so—which probably they could by employing men who understand the business. There are men in the northern and western states who could find employment here at this business, and I presume there are those who could make money at the setting and rearing of hedges until a year or two old, or until a complete stand is ensured and sufficiently grown to be given over to less experienced hands to work and manage. Scores of men in this country have attempted to make fences of this kind, and have failed for the want of experienced help and management.

That fine hedges can be made here is a certainty, for I have seen here and there short pieces of fence of this kind that was as good as need be for any purpose. In fact, I have seen short pieces of hedgerow that grew up and came to a good fence in the open prairie, without any attention after being set, and poorly cultivated and trimmed only two seasons and then left for thousands of cattle to tramp and interfere with, if they were so disposed, every day. And, again, I have seen good, well grown hedge-fence from other kinds of plants in this country, but I am sure the Osage, with good management, is much superior to anything that can be used in the country for the purpose of hedging. Young men who understand the setting, pruning and correct management of hedges could go to Texas and soon accumulate enough to start the stock business.

The producing of honey can nowhere be made more profitable than in western Texas, from the fact that bees can work almost the whole year, and that the prairies and forests are covered in such profusion with flowers and blossoms, and the different material from which they extract their honey. They are not liable to freeze here, and are perfectly secure from all the dangers and difficulties of winter in colder countries. It is said that bees do not do so well here when kept in the open prairies as when kept near the bottoms along the rivers, or in the prairie forests, where the winds do not interfere with their flight to and from their honey flowers, or the material from which they extract the sweet.

If those men whom I have known at this business in other parts were here, taking one-half of their accustomed pains at the business, they could soon make enough money to enter largely into the stock business. The poor man, without a dollar, can secure bees from the forests and fill as many hives as he pleases; and a person with a little spare means could build those shelved bee-houses which have long since proved so fine and profitable, and which prevent the trouble and losses from swarming.

The writer himself is a poor man, and, as few as his virtues may be, he can say that he has been good to the poor man. When I tell him that I have been in all the grand divisions of the globe but one, and have seen much of the world, and that western Texas is the finest and most promising country of any I have seen for the poor man, I am not trying to deceive him.

When I tell him of a country where a few yards of strong cloth will make him a house that will do him until he can make something to enable him to erect a better one; of a country where he can pasture his cow, his sheep, his horse, his hogs, etc., all for nothing; where snow never comes; where the soil never freezes; where the man who has the least bit of money-making tact is almost sure to get rich or become independent; in short, of a country where he can live and ask but little or no odds of those who gripe their money; where he can shoulder his gun and go forth with a certainty of returning with meat for his family; where the lustrous rays of the sun, moon and stars are cheering to his heart and afford him a greater delight than do the cartwheel dollars given to the sordid and miserly man; where he can contemplate the bounties and beauties of nature with enjoyment, regardless of the dazzle and grandeur of fashionable life and empty show; and where, as I have said before, a man can be a man—not a dog…

I say, when I tell the poor man all this, and that this is the place to live, I know he will not think I am trying to deceive him. When I tell him that I have swung the ax, the cradle, the scythe, the fork, the hoe, the spade and have stood by the plow, and have toiled in the field year after year; have clipped the fleece from thousands of the woolly tribe, and have done almost everything that the laboring man does in northern countries—he will admit that I ought to know what is good for the poor man; that I ought to be a competent judge of a country's advantages and suitableness for him; that I ought to feel an interest in his well-being, since I have realized and know so well his task—since I have realized and know so well the stiffened joints that sometimes almost refuse to move when he rises at the break of day to return to the field to work on and lend his entire might in the production of that which clothes and feeds the rich, which warms and nourishes the sick, and showers such blessedness upon man. Yes, he will admit that I ought to know all this and undoubtedly do know it and am telling the truth.

So, farewell poor man! With these and a thousand other things I might tell you, until I meet you in western Texas. But, you may say, hold for a moment, and tell us why you are so poor, after being so long in this fine country, where you say it is so easy to become rich. This you have a right to ask, and it is my duty to answer you, which I can do in a few words.

The writer happens to be one of those whose disposition it was to climb Fortune's Ladder by skipping too many rounds for the purpose of going too rapidly up. Consequently, losing his hold, he came prostrate to the ground. And when coming to Texas, after trying his fortunes in different parts of the world, instead of improving little advantages that lay before him, instead of buying a few cattle, a small flock of sheep or something of the kind, and going gradually but surely up, he misplaced his confidence and undertook again to climb the quick way to a fortune. This time, the ladder, whose ascent he would have handsomely made, proved deceitful and rotten.

My readers may not all understand this, but there are, perhaps, a few who will, and that will answer the purpose. When arriving in Texas, the poor man should not be discouraged and suffer himself to think that there is nothing here that he can do. He must remember that he is from the same country where old Texans came from, and he must pitch in at the best thing that offers. Let Texans know that he is neither "pepper or salt," "sponge cake or dough," but rough-and-ready for the work of the country. Pitch in at any price for a while, and let 'em know what you are good for, and you will soon be all right.

Teachers, Ministers, Etc.

Common school teaching in western Texas is that to which I would call the attention of the poor man of education. The recent census of Texas shows a population of nearly half a million people. Consequently the pupilage of the state is large and rapidly increasing, which makes it an inviting country to the teacher—a broad, expanding field for this fraternity. The school teacher can here put his earnings into stock, and have it attended to as well as though he gave it his personal attention. During vacation, he can ride among it and satisfy himself in regard to the business or management of his cattle, sheep or whatever kind of stock he may have his money invested in. By laying out his surplus money in this way, he would soon have an income from his cattle that would equal his salary for teaching, and before many years, with prudent management, he could take up his abode in some good grass region, with his hundreds and possibly thousands of cattle around him.

I can imagine nothing more encouraging to the patient and perplexed school teacher than to have an opportunity of investing his little surplus or savings in a way that it shall be rapidly compounding, or in a way to get the free use of a great natural pasture, upon which the increase of his stock will carry him to wealth.

Let me ask the district teacher how he would like to have his brand upon cattle that are grazing on the commons, and all about him, the keeping of which might not cost him a cent, and the marking and branding of the increase of which he could pay for by the head, or give a portion of it for the branding, etc.—among which he might ride at leisure times upon his pony that might always be close by, staked out, or hoppled upon grass near the schoolhouse. All this the teacher realizes in this country, and the stock boys are glad to see him start his fortune and reap the benefits of the yielding earth in common with them. It is not unusual here to see a half-dozen or more horses staked upon grass near the schoolhouse. What is more encouraging and satisfactory to the friends of education than to see those horses, coming down from the outskirts of the district, loaded down with healthy and happy children on the way to school?

If the teacher prefers it, his investment might be in a flock of sheep or a herd of mares, to be kept by someone not far from his place of teaching. In several instances, I have known this course to be pursued by teachers who would at times ride with the stock boys, help drive and brand, and become acquainted with the business. And I presume, should I make a little inquiry, I could find out many here with fine properties who, in the commencement, pursued this same course.

It is not long since I knew two young men, brothers, to purchase five hundred head of cattle, paying a portion down and depending upon their salary as teachers and the few beeves they might have the next year to pay their notes for the balance.

Of course there are many high schools in Texas, good teachers of which are appreciated and well rewarded, and there is a disposition in the south to depend more upon itself for education of its children. They prefer having schools near home rather than be compelled to send their sons and daughters north for an education. Their reasons for this are many and sound, and this determination will be the origin of many private common and high schools and colleges which, of course, will multiply the demand for teachers of all kinds. Teachers of music can nowhere do better than here. Ornamental education of the different kinds is sought and well paid for in this country.

The school fund of Texas is very large, and there is always a great surplus for educational purposes in the treasury of the state. Our system of schools is very popular, and we boast of affording superior means of education to the many. In fact, our school system is claimed to be modelled after and an improvement upon the school systems of the older states. This, I suppose, arises from there being people in Texas from all the different states of the Union, and the manifest disposition to introduce here anything that is good from the older states. The voice of New England is heard in Texas.

It is needless to say that the churches and religious societies in Texas exercise a powerful influence upon the morals of the country and lead many into the holy bonds of the Christian religion. Churches and places of worship are quite numerous and rapidly increasing. Our clergy are proud of their country and labor devoutly and with zeal in their cause. Our benevolent societies and institutions speak well for the hearts of our people.

It is not unusual to see the brand or mark of our ministers upon the cattle of our prairies, and many of them have fine stocks of different kind of animals, which frequently have accumulated from little beginnings. The generous stock boy sometimes by way of compliment and as a commendable act, puts his minister's mark and brand upon one of his heifer calves. Of course all would look out for the minister's cattle; and, as he is so generously remembered, his stock is ever on the increase. As a natural consequence, it becomes large and valuable, the proceeds of which afford him an easy support for his family and enables him to educate his children as he would wish at home or abroad. One minister I know is not only the shepherd of his flock in the church but the shepherd of his flock on the prairie—a beautiful flock of the woolly tribe. Often when working with a crowd of stock boys, I have known some one or more of them to brand all the motherless calves they might find on the prairie, and now and then a yearling that has left its mother, to some little orphan girl or boy.

I have nowhere mentioned that when a stock of cattle is neglected, and the calves get to be yearlings and leave their mothers before being branded, they are liable to be taken up and branded by any stock grower who may find them. This is perfectly right and will make stock growers look after their business in time.

LABOR IN DEMAND

From the beginning of this work, it has been the writer's intention to note down every description of labor for which there is a demand in western Texas, also all the different vocations and branches of business in which men are engaged. But the want of space and time, and the writer's neglect to make a memorandum of his observations from time to time upon these points, renders it quite impossible for him to do so.

As a general thing, the different branches of business of a country indicate the kinds of labor wanted, and by reading this work, the laborer of any kind would be pretty well qualified to judge for himself whether western Texas is the place for him or not, even though I should say nothing more on the subject. I will, however, after adding several branches of business to those already mentioned, put down several kinds of labor for which it occurs to me there is a good demand in western Texas:

Lumbermen with portable sawmills could nowhere do better than in the timbered sections of the country. Good hotels and boarding houses do remarkably well here. Good, thorough farming, independent of the stock business, pays exceedingly well—halfway loose farming does not pay, for in this case thrifty weeds are sure to take the field. The livery business nowhere does better than in Texas—the prairie grass is free to the liveryman, which he can have merely for the price of cutting and drawing his stable. The sale of northern wagons and carriages of all kinds is very extensive.

It is enough to say that the pursuits of the towns generally of any civilized country, such as butchering, draying, blacksmithing and general mechanical business pay well here. Laboring men about our coast towns and wharves generally get good pay for their services. Stage drivers get from $20 to $50 per month, and many of them soon get into the stock business or become speculators in horses, mules, etc. Horse, mule and ox-teamsters get from $20 to $35 per month, and can accumulate by being economical and investing their savings in stock of some kind.

Stock drivers get from $20 to $40 per month, and soon begin to accumulate property in the way of stock, etc. Beef buyers and drivers get from $40 to $125 per month, and with economy and good management can soon get rich. Northern and European sheep-herders or shepherds get from $15 to $30 per month, and can get ahead by putting their savings in sheep. Sheep-shearers get from four to six cents per head for shearing. Managers of large flocks of sheep get from $30 to $50 per month, and can get rich by investing their surplus means in sheep.

Plantation overseers get from $300 to $1,000 per year, and many of them work their way up in the world. Clerks in mercantile and other establishments get from $250 to $2,500 per year. Laborers at general farming get from $15 to $25 per month. Ditching is done by the rod and pays the laborer well. There is a great want of servant girls in western Texas, and any number of servant girls could get good situations at fine prices.

In conclusion, thousands of apt and handy mechanical and general laborers could find enough to do in Texas and would find it profitable to go there. Although railroads have progressed rather slowly thus far in western Texas, they are sure to be built and thread our country in all directions, and this will create a renewed demand for labor of all kinds. The constant emigration of wealthy and enterprising people is sure to bring about the extensive construction of railroads, which will open a wide field for all kinds of business and labor. It is thought by many intelligent men that a railroad reaching the Pacific will someday have an eastern terminus on the gulf coast of Texas. It would seem that such might be the case.

TEXAS

INDIANS, SNAKES & INSECTS

Indians, snakes, poisonous insets and mosquitoes are supposed by some in the north to be the great plagues of Texas. Along the bordering settlements of Texas, at times, Indians are troublesome. But new settlers seldom go where they would be likely to be disturbed by them. In all the writer's travels, for years in western Texas, he has never seen an Indian here unless it was one who passed for a Mexican. The opinion abroad in regard to Indians is entirely wrong. They are only troublesome at those points where newcomers would not be likely to settle.

Snakes and poisonous insects are quite plentiful in parts of western Texas, but where cattle and hogs or stock of any kind are kept, they soon disappear. Hogs devour them, stock boys shoot them, and they give way to civilization. During the writer's six years of residence and travels in western Texas, he has never known a single individual bitten by a snake or a poisonous insect. That they do bite and that some of them are deadly poisonous, without ready relief, I suppose is certain. But men are so seldom bitten by them, and they so seldom die when bitten, that I can see no good reason why there should be so much noise made about them.

The people of Texas think no more of snakes and insects, and give themselves no more uneasiness about them than do the people in other countries. I suppose the awful reports sent abroad from Texas about them are from newcomers, and women principally, who no doubt have a perfect horror for and loathe the snake or poisonous insect, as they picture them by accident under their hoops or in some way in their imaginations over and around their heads.

Mosquitoes along the coast of Texas and as far back as the level country extends, are at times quite plentiful and to newcomers somewhat annoying. But they too give way to civilization. As the country is settled and is trampled by stock, they diminish in number. The settler who locates upon an elevated dry spot, at a reasonable distance from ponds of water or dense forests, need never be troubled with mosquitoes about his dwelling or anywhere within its vicinity if he has stock around him to eat down the grass and tramp the soil.

Hard Times & Money Matters

It is an interesting fact that what is known in other states as hard times is never known in Texas. To be sure, men get cramped in business, but it is from their own bad management or some local cause, and not from a general stagnation in the business of the country. Although Texas imports largely of those articles that she will eventually export, which must necessarily act as a drain upon the country, money seems always to be plentiful and business uniformly active. I might give some reasons for this, but it is enough that it is so.

Texas Almanac

Persons wishing to know more of western Texas than this work affords them can, by addressing Mr. Richardson of the *Galveston News*, obtain an annual, statistical and general history of the state called *The Texas Almanac*, together with a cheap map of the state.

This almanac gives a description of every county in the state, stating its agricultural capacity and price of lands, the amount, price and kind of stock raised in each county; the condition of schools, churches, railroads; and, in fact, a large amount of useful information pertaining to the state and its prospects. For general and correct information in regard to Texas, no one could do better than to send for the *Almanac*, which is issued at the beginning of every year, and should be in the hands of everyone who thinks of going to Texas, or wants reliable information in regard to it.

Who & How to Go to Western Texas

To make a comfortable and respectable home in western Texas, there is nothing like as much required as in the north. A hundred and one things are used, indoors and out at the north, that are not really needed here. Therefore, whoever starts out for this part of the world with a family, should take such, only, of their household things, farming utensils—bulky, heavy and cumbersome articles—as they may think will be really necessary here. It does not pay to move old, worn out furniture, hardware and the like, such a distance. So, I would say to those who go to Texas—reduce your stock of movables as reasonably as you can, and depend upon buying in Texas, when you can find what your really need here.

I do not mean by this that you should leave all your woolens and warm clothes behind, or your nice, nearly new cook stove, or your favorite bureau and mirror; but simply that you should not take such trashy, half worn-out stuff as is not worth moving. Get what you can for such things as you will not need in Texas, and bring it here in your pocket to invest in stock and lands here. There are plenty of families in the north who have enough about them that they would not need in Texas, which, if sold, would buy cattle, mares, sheep and stock enough of some kind here to give a person a respectable and encouraging start in the world. There are thousands of families in the north whose circumstances are embarrassed and who are constantly pushed to get along, whose moveable property, in and about their houses, including chinaware, etc., over and above what would be needed in Texas. If turned into cash, it would buy in this country a beautiful flock of sheep, a fine tract of land and erect a comfortable dwelling upon it.

Think of the change, you who feel pressed and poor, yet with enough around you, more than you need, which, if turned into money, would make you rich in the most delightful climate and beautiful country. How many fathers there are, who, if they knew how far their little properties would go in Texas, would sell out and find their way here immediately. How many there are who will tug along the path of life and leave their children little or nothing, while others will go to western Texas, invest the little they have and live out an easy decline of years, leaving all their children a fine property. If the sons and daughters of many a family in those cold and snowy regions only knew how they would be delighted here upon the evergreen sod of Texas, and how wealthy they might get here, they would give the father no peace until he should pull up stakes and start for the Lone Star Land.

I said this was a good country for the man who would retrench, or who is unable to keep pace with the prodigality of the times where he may be living. Are you the man whose poverty is groaning under the extravagance of empty show, and whose course must be changed in order to avoid the unreasonable demands of inexorable and unfeeling fashion, and its ultimate disastrous effects? To be plain, are you the man who must get away from where you are, in order to curtail your expenses and place your sinking means in some country where they will afford you a better income, and enable you to meet the necessities and requirements of a large and growing family, without being constantly harassed by creditors, with the prospect of finally surrendering your all in keep up appearances in the circle in which your family is moving?

If so, sell out, settle up, get ready as best you can and come to western Texas. You can here provide yourself with the best of lands, for a trifling amount, and build you a spacious and comfortable house and all necessary improvements, with but a small outlay of money, and be as free from the follies and whims of fashionable life as you please. Yet you need not be entirely secluded and shut out from good society, unless in your selection of a home you so prefer to be.

By looking through this country, anyone who is in pursuit of a home, whether a professional man, merchant or mechanic, can find the place that would suit him, unless he might be more nice than wise, and look for that which no new country ever affords. The farmer or stock grower can find a section that is good for the raising of cattle, horses, sheep and hogs, without his producing, if he prefers not, anything for them to eat—a section that is good for raising wheat, corn and almost every vegetable that grows, various delicious fruits, which need not be far away from a good school. But on account of schools, society, religious and other matters to which I shall now devote a few pages, I shall advise a way of coming and settling in western Texas that I believe will make the foreigner overjoyed with the country, and result most gloriously for himself and posterity.

There is one thing about settling in new countries to which I would call the particular attention of the emigrant, and that is the mistaken idea that friends are to be found in distant lands to fill the place of those of early life. I care not how beautiful the country, or what advantages it may offer to the settler, or what kind of people and society he may there find—he will not be as well satisfied with and as well contented in such a country, by leaving all his relatives and friends of early life behind, as he would be if more or less of such friends or relatives were settlers with him.

A person may start out and travel over the world alone, and not feel his separation from youthful friends so much when he carries with him the idea of someday returning to mingle again with those whom he became attached in early life, memories of whom he cherishes in connection with thoughts and comings of the future. But when he goes alone, or even with his family, to settle down in some new country, among strangers entirely, he is too apt to feel a loss of that which he can in no way make up. Time may roll on, and years may go by, and he may return to the home of his birth which, although he loves, he would not exchange for that of his adoption. Yet he asks, where are the friends of my early years with whom I used to caper and play? Where are the cheerful and intelligent girls, in whose company I once loved to be? Where are the good old people of my neighborhood? Where are the friends of my youth, those friends of my happy days, such as I have never since found, the loss of whom has left a vacancy in my heart, a dreaminess in my life? How often has this settler exclaimed, "O! for one good old-fashioned neighbor and friend!"

Therefore, I would not ask a person to leave behind that cluster of endearments that have forever entwined around his heart, and all his relatives and friends of old, to take up an abode in a distant country among strangers. Paradise itself, with all its enchantments, could not withstand the encroachments of feelings occasioned by the thoughts of leaving all the friends of early life to accept an abode even in her most favored spot. Judging the feelings of others in this respect by my own, I am always glad to see parties of emigrants linked together not only by feelings of interest in and dependence upon each other, but by feelings, if possible of solid friendship and kindred love. Such kind of emigration is worthwhile.

But to hear a person relate that he has left everything that was dear to him—friends, relatives and all—perhaps across the wide deep, or thousands of miles away, it makes me sad. Although he may have come to this fine country where his prospects may be ever so good, it is sad to hear his tale of loneliness. A person of talent or shining abilities, of property in abundance, or one who is gifted in the way of getting into society and can always make himself agreeable to others, or, in other words, one who can adapt himself to circumstances, might not feel the force of these remarks as would those who are without those qualifications. As a general thing, the lonely emigrant or family who settles in a distant country among strangers, forever mourn their complete separation from relatives and friends of early life.

There is something of still greater importance that the lonely emigrant is too apt to feel the want of—that is a way to educate his children properly and give them the advantages of moral and religious examples and instruction. Strangers in new countries, and in Texas particularly, do not seem inclined to settle near each other, but very often locate as far apart as possible for the purpose of gratifying their avarice and having a vast range for their stock, where perhaps one spear of grass in a million is not consumed by their cattle. But relatives, friends, old acquaintances and even those who go from the same section of any country to a new one, are apt to settle with a view to neighborhood schools, society, religious advantages and many other inestimable interests. And it is well known that when people settle in this way, those feelings that have so long existed between them become strengthened. They learn to regard each other as brothers, and their intimacy is source of such happiness to them as they could not have attained by settling separately and among strangers.

I would not be understood that the thousands of men and families who go alone into a new country are never happy and contented, but that when they do so, they go without a peculiar source of happiness, for which they may ever after look in vain. I know the world is full of reputation, wealth and steadfast friendship that took root among strangers in strange lands, but who knows of the sadness occasioned by the thoughts of this separation from everyone and all that was once dear to the possessor? Who knows what it has cost him to do without a single friend of old?

I have noticed that settling among strangers is apt to create a sense of indifference in regard to the education of children. Parents are too apt to say that, "My children are not going to such a school, among strangers, and with such and such children. I will wait for a better opportunity." And from the simple fact of being among strangers, thousands of children are neglected during that period of their lives when all possible attention should be given to their improvement.

For these and many other reasons the writer is strongly in favor of party or company emigration and neighborhood settlements. Certainly it is that when people of like peculiarities of society and life go to a new country and take up near to each other, they are better qualified to appreciate and understand the excellencies of such country than would the lonely emigrant, whose mind from the nature of things would be occupied with that which renders daylight in appearance too much like dark. Therefore come not alone but bring with you your old friends, and western Texas will prove as near to that paradise of poetic and Bible renown as would any country in the wide world. Come in a way that memories of the past and cherished things of yore will not weaken your judgment and ability to appreciate your present surroundings. Bring the spirit with the body and you can then understand the beauties of our land.

I do not mean to say that new countries should be filled up by people all from the same country, but that a few intimate friends or families, or people from the same sections, of the same conditions of life, would be happier settled near to each other than among strangers in a new country.

I said that people in Texas are inclined to settle away from each other. This is the case to a great extent with the stock growers of western Texas, and the reason for their so doing, aside from their separation from their friends of old and forgetfulness of duty and disinclination to society, is that their minds are so glutted with the superabundance of the country. They are constantly gathering up their cattle from among other peoples' stock, where perhaps the tops of the grass are only eaten off, but not one-hundredth part of it consumed, and pushing them far out upon ranges where nothing but the Indian, wild horse, deer and the like have for past ages made any pretension to possession or consumption of the inexhaustible production. They can have it all to themselves and as little as possible to do, and too often smother their duty to their children, society and almost everything that belongs to man, but property and money, or idleness and ease. To be sure, they get upon their horses and go several trips a year for the purpose of branding their increase, amounting in all, to three or four months' work, such as it is, and the balance of the year is too often most woefully squandered and unprofitably passed away.

I would not have the reader understand that these remarks set forth the habits and condition of all stock growing communities in western Texas, but that there is a great disposition among stock growers to settle far away from each other, which deprives them of the advantages of schools and society, the cause of which, to a great extent, is the want of intimate acquaintance, early attachments and like peculiarities of life and raising.

There are plenty of stock growers in western Texas, living some distance from a neighborhood or school, who mount their children upon horses and they gallop away to school. But some are too indifferent in regard to schooling for this, and others live too far away to send their children to school. Many send their children to towns to board and attend school, or to a boarding school, but this occurs too often after their younger days are neglected.

Now I will suggest a way of emigrating to western Texas that would result most gloriously to the settler and his posterity, and I can see no reason why it should not be practicable. If I were a resident of an old, densely populated country, where perhaps I could find nothing profitable for me to do, and should take it into my head to go to western Texas, and should be pleased with it and prefer to make it my home, what objection could there be to my returning and interesting with me a half dozen of my former friends, in a way to be satisfied and contented? And for our convenience and benefit, what could be more becoming than to interest with us several mechanics of different kinds—say, a blacksmith, shoemaker, tinner, gunsmith, wagon maker, cooper, harness maker, tailor, millwright, carpenter and possibly other mechanics; a merchant and physician, an attorney, a good schoolmaster, a printer and a minister of the Gospel, and as many of the laboring class as the company might see fit to induce join them?

After getting together and appointing the right kind of prudent managers, should there be found a want of funds to enable the enterprise to go forward, how would it do for the managers of such a company to communicate to some gentleman of ability, "that a number of respectable citizens and families [of whatever place it might be] have determined to go in a company to western Texas for the purpose of settling there as farmer, stock growers, mechanics, merchants, etc., to improve the country and be useful to themselves and mankind, that there are poor people in the company, and that it is in want of means to enable it to go all together and settle down in a way to get along and be satisfied as pioneers, and we as their managers are authorized and instructed to request you to address the community on the subject of emigration, for the purpose of forwarding this work. Could you make it convenient to interest, with your influence and talent, the people of the community in this subject, and let this company feel their obligation and gratitude to you for having rendered the assistance they so much needed," etc.?

Could not something of this kind be done, and collections taken up in this or a similar way, with perfect propriety, to enable poor people to emigrate with those who are better off? Would this not be the way to advance the work of civilization, to carry light into benighted Mexico?

Would it not be somewhat in imitation of the great Austins, the colonizers and fathers of Texas? Were the noble Austins the last of old Connecticut's sons who will brave the dangers of the frontier and open up new fields of civilization, and call out the usefulness of men?

Would not such a man of mind be proud to sound his voice in such a cause? Would not his soul fill with eloquence when considering such a subject? Would not the people of any community be eager to listen to an able speech and orator upon such an occasion? And would they mind a quarter or a small admittance to the hearing of such an address? Methinks I hear the voices of Everett, eloquently discoursing upon the expanding importance of this great land, carefully lifting the veil of the future, and prudently directing the emigrant upon his course. Methinks I hear the voice of a great and wise man commending this laudable scheme of emigration. Methinks I see a group of intelligent men earnestly considering the prospects of western Texas and mineral Mexico, and they decide to come and live in these money-making and happy regions.

The spirit of Washington pervades our whole country, and the same God that watched over the battles that won the liberties of America is still watching over this expanding confederacy and, it is to be hoped, will never abandon it to an unhappy fate.

A very good way for poor people to get to Texas is to get in with those who are taking stock there, either by land or by water. Go along and help drive or attend to stock, and so work your way to the country. There are many going in this way.

Different Routes to Texas

Those who get to Texas from the southern states generally go by the way of New Orleans, and across the Gulf of Mexico, or by Red River to Shreveport, or some point on this river where they leave it and cross the country by land, both of which routes to Texas are much traveled.

There are different routes for those who go from the north, east and west. There is a regular line of packets running between Boston and Galveston, which affords a quick, comfortable and reasonable passage. There is also a good line of packet ships from New York to Galveston. There is regular steamship communication between Galveston and the ports of western Texas—Indianola or Lavaca and Brazos Santiago. There is also a regular line of a smaller class of ships running between New York, Indianola and Lavaca. There are schooners running from Boston and New York, and vessels from Philadelphia to Galveston and western Texas. There is a quick steamship route between New York and New Orleans. Many northern people take this route to Texas, and some take what is called the southern route, by way of Washington, on to New Orleans and across the gulf. The last two routes are principally for passengers, and not so much for freight and luggage.

Others go the western route and down the Mississippi River, either to New Orleans or the mouth of Red River, and thence as stated above. The price of passage from Cairo and St. Louis to New Orleans is from $12 to $30, according to circumstances. The distance from Cairo to New Orleans is about a thousand miles. The price of passage across the gulf by steamship, from New Orleans to Galveston, is invariably $15, and $20 to Powder Horn or Indianola, and I think about $20 to Brazos Santiago.

There are two routes from New Orleans to the ports of Galveston and Indianola—the inside and outside route. The inside route takes you over about seventy-five miles of railroad, extending from New Orleans to Burwix Bay, thence by steamship across the gulf. The outside route is performed entirely by steamships, down the Mississippi River to its mouth, thence across the gulf. Some prefer one route, and some the other. The price of passage over either is the same. There is no change by the outside route, and the ships are larger than those of the inside route. The Burwix Bay or inside route is several hours the quickest. It will not be long before the Burwix Bay railroad will be extended on to Galveston and central Texas.

Those who cross the Mississippi River at St. Louis, above or below, and go by land to Texas, I am unable to direct, further than to say that there are noted crossings at all the large rivers and good routes are to be found, of which anyone can learn anywhere along the Mississippi. There are regular thoroughfares across Missouri, Arkansas and the Indian Territory to Texas, which are perfectly safe, and along which animals get all the grazing they need, free of expense. There are many going these overland routes on through northern to western and southern Texas. People going these routes with wagon and horses, or oxen, can lay in and occasionally buy provisions, camp out and travel very cheap. Sheep and all kinds of stock, I learn, are being largely driven across to Texas from the western states. The higher up the route, the better it is considered for stock driving, as grass is more plentiful, there is less timber, fewer settlements, and crossings of rivers are more shallow and easily gotten over. The roads higher up the country are said to be better. Some who drive cattle from Texas north go entirely around Missouri, by going through the Indian Territory, Kansas, the corner of Nebraska, into Iowa and so on. Some drovers cut across the upper or northern portion of Missouri.

It is said that forty thousand sheep, on their way to Texas, passed through a small town last year (1859) on one of these overland routes. Probably not less than a hundred thousand sheep will be taken into Texas, from different points north of Texas this year (1860). Let them go. That is the place for them. I see it is stated that two hundred and fifty thousand sheep were brought out of Mexico in 1859, but this must be a mistake.

Many are the wool growers who are now making their way to western Texas. The man who can make a raise of a pair of horses or oxen, and a wagon, together with a small flock of sheep, had better join some friend, mix flocks and find this way across the country to the sheep region of Texas, that is if he would like to make a fortune at the business. The man of but little means could, if he chose, take his family in this way, and drive his sheep along with but a trifling expense. He could take in his wagon nearly all that a small family would require at first, to keep house within western Texas, or that portion of it where snow is seldom seem; where cellars, filled with vegetables for the winter are not required; where a little open shanty will do until a better one, or a good house, can be built.

In going these overland routes to western Texas, a person may pass through the wheat region of the country, and no doubt be pleased with northern Texas, and would settle down before reaching their destination.

How to Find a Location

When arriving in western Texas from across the Gulf of Mexico or any point at sea, unless a person may have friends in the country and knows to what point in the interior he is going, the questions will naturally rise with him: What shall I do? Where shall I go to find a location? How am I to satisfy myself in regard to this country? And, possibly, he may ask, what am I to do with my family that they may be comfortable and yet not bring a heavy expense upon me while I am engaged in exploring the country to find a suitable location for my business and future abode, where I could make it profitable and should be satisfied to live?

A family or a number of families arriving upon the coast of Texas in this way, to accomplish these objects in a judicious and economical manner should, if their intention is not to live upon the coast, go a little into the interior, where the unoccupied country is rolling and dry, where there is plenty of shade, wood and water. Camping upon some pleasant stream is a way to get along comfortably with perhaps not more than one-fourth the expense of boarding at a hotel. I say after accomplishing this in a judicious way, the gentlemen of the families or such of them as may be best qualified and are agreed upon the purpose, should secure good, substantial and gentle ponies, saddles and traveling equipage, coffee and coffee pots, tin cups, sugar, dried meat, bacon, hard biscuit or flower, and possibly a

Mexican who can talk your language to pack and lead your mule or horse, to look after camping fixtures and do the cooking. After the first is accomplished and the last secured, the gentlemen agreed upon, with a map of the state in their pockets and a memorandum of what has been learned at hand, should start out and give the country a good looking over.

Those who intend wool growing as their business should, of course, go to the sheep raising sections, examine the ranges and different flocks of sheep, talk with shepherds and proprietors, and be not afraid to push a little further on. Go up this creek. Cross over down that creek. Make all necessary inquiry and when a good-looking unoccupied range is found, where lands are for sale, satisfy yourselves in regard to its suitableness for you—learn who has the title, for the owner may be a hundred miles away. Learn as much of the title as possible and be careful that you are not imposed upon by those who would not have you buy as well as those who might be anxious to sell. Continue your work in this way until you have found the place you want, and then either buy, lease or contract, or make arrangements for a home upon it according to circumstances as I have repeatedly directed.

In looking out for ranges of any kind, whether for cattle, horses or sheep, a good hog range will never come amiss. But if sheep raising is the principal thing in view, do not surrender its claim to that of hogs or anything else. If a range for all kinds of stock is desired, a small section of the range at least should be quite rolling and dry. This, of course, should be for the sheep. For horses, the country must be moderately rolling, and if it is hilly and mountainous, even, they will do well. For cattle, or Texas cattle at least, it matters not so much. I believe the stock growers of western Texas, who make the most money at the cattle business, have their stock upon ranges near the coast. But I would say to those who take Durham stock or northern cattle to Texas, not to attempt to keep them too near the coast on account of occasional wet seasons and mosquitoes.

Now, if the reader will carefully observe what I have said throughout this work, although some of it may seem of but little consequence, he will, if he settles in western Texas, be able to steer clear of many bad bargains and blunders that he might otherwise make.

There is one thing in regard to building spots in the lowlands of Texas to which I would call particular attention. When I speak of lowlands, I do not wish to be understood as meaning wet or swampy lands, but the lands along the coast which are generally wide open prairies, much inclined to a level, but here and there slightly and sometimes considerably rolling; also at greater or less distances, divided with streams and rivers, along with there is generally some timber. Now this section along the coast, which is an immense cattle range, has heretofore been considered, its river bottoms excepted, the poorest portion of the state unless we except also the desolate plain in the far west. But this is a mistake. These lowlands will yet be found to be the most valuable in the state.

I have no room to give my reasons why, but it is enough to say that they are rich, inexhaustibly rich, and are sure in time to be appreciated. What I am now at is this: when a person locates himself upon these coast lands, he should aim to secure a building spot that is somewhat sloping or rolling, and where if possible a sea breeze can have a fair sweep—not behind a thicket of timber or underbrush, but an open spot where the water will readily run off. I do not mean by this a *hill*, but a sloping or rolling spot. If these directions are observed and cistern water is used, after one or two years' occupation and acclimation is experienced, sickness and mosquitoes will be strangers to your house and surroundings. This subject of finding a location in western Texas, if rightly treated, would be very lengthy. I will, therefore, drop it with the few suggestions I have made.

Slavery

Although I have seen a great deal of slavery as it exists in the south, I have not acquired that practical knowledge of it that I might have done. Much of my time in Texas has been spent with that class of people who have but few and, many of them, no slaves at all, and in sections where slavery is not considered profitable. But from the acquaintances I have with this institution, I should say let it alone. Were it left entirely alone, in the opinion of the writer, labor in the United States would find its true level. It is easy enough to talk about the evils of slavery, and to agitate with a view to its abolishment, and were there but a few thousand slaves in the country, they could be set free and no great calamity would arise from it. But when men talk about setting the millions of negroes of the south free, it is time we are talking about a great national calamity. What could be done with the slaves if liberated? Should they be left to work as free laborers in the south?

Let me say to those of this opinion that when in Africa, not long after the liberation of slaves at Cape Colony, the writer put up at a hotel kept by an intelligent English lady who had a great deal of difficulty to get along with her household affairs in consequence of her slaves having been set free.

She said that since the slaves of the colony had been liberated, there was no dependence to be placed in them; that they were a nuisance and would not work enough to keep from starving, and only at such times as they pleased. Sometimes the lady and her daughter would do the work of their large house, and again they could hire one of the liberated slaves and get along very well, until the hireling would get drunk or take a notion to lounge in the street. I saw plenty of idle ones in gangs of five, ten, fifteen and twenty, all over town, and they were saucy and impudent enough.

The lady told me that a farmer of her acquaintance, after giving up his slaves, attempted to hire them and get along in this way. When his crop of oats was ripe, he went for his help, and found two of his men under and orange tree with a jug of whiskey. He made known his errand and their answer was that they did not want to work. After long persuasion and extravagant offers, they finally agreed to come and cut the farmer's oats, after they had eaten all the oranges off the tree and drank up their jug of whiskey.

What an orange and whiskey orchard it would make of the south in exchange for its most beautiful cotton fields, if her slaves were liberated and left free in the country!

Now let us suppose that the slaves of the south were set free, and sent away to take care of themselves. My father once had a negro. They called him Jack. When sent out for a basket of chips, he would fall asleep while picking them up. When climbing a fence, he would sometimes get to the top, go to sleep and fall down. Jack wasn't profitable without a whip behind him, and *he* wasn't *niggers* enough to pay for an overseer. For this reason he was either given away or sent off to shirk for himself. At all events, the story runs that he went to riding horse on the canal, got asleep, fell in and drowned, of course. Now if the great mass of negroes in the south were sent to shirk for themselves, in any part of the world, would they not fall into the canal, or what would be equally bad, get asleep while picking up their chips?

If slavery can live in contact with free labor, let it live and *peaceably* too. Slaves are human beings; why turn them upon the world to starve? Go to the cotton and sugar fields of the south and look about you. Ask yourselves where could these millions of negroes be so happy, and where so useful to mankind? Then consider and treat things according to circumstances, and think and talk of things that are practicable.

Relaxation of Energy

I have somewhere said that the greatest objection in this country is the easy means of support that it affords to men, which has a tendency to beget habits of idleness. Some attribute the laziness of men in Texas to the peculiar effects of the climate upon the system. But I believe, and ought to say, I know that such is not the case.

The most thrifty and robust young men I ever saw were the Anglo or white natives of Australia, and if ever there were lazier white men than the Australian stock growers were, then I have not seen them in my travels. The health of Australia is better than that of Texas for one reason—its native growth of vegetation is not equal to one-third of fourth of that of Texas, and its trees shed a thin scale of their bark instead of a mass of leaves. Consequently there is no vast decay of vegetable matter as in Texas. Therefore, other things equal, the air of Australia must be purer than that of Texas, when first settled. But when the country around the settler almost anywhere in western Texas, is relieved of its immense annual decay of vegetation, there is then, in the opinion of the writer, no real cause for relaxation of energy in the settler.

Pure, wholesome air and water will never take away a person's energy, nor render him unfit to labor in any country; neither will it make him lazy. But a country that will afford a good living and make a man rich, without his storing up a little for winter, neither for man nor beast, and is otherwise peculiarly easy for the occupant, would very naturally make him lazy. To excuse himself, it does very well to say that a white man can't work in this climate. The truth is, men can work in Texas as well as in any country, and I sincerely believe that most of the time it is the most delightful climate to labor in that I have ever known, unless a person suffers himself to think that he is in a country where there is not much labor required of him, and indulges himself too much in this way until, perhaps he thinks he can't labor.

I am well acquainted with foreigners here who can labor the year through at the cultivation of the soil, and they tell me that it is all nonsense to say that a man can't labor in Texas. That there is no necessity of his working so hard is very true; but that he must for this reason become lazy and idle is absurd. The industrious man who comes here from any country can continue his industry as pleasantly and more profitably than in almost any other country.

Mexico Hopelessly Fallen

I have stated that the disturbance by Cortinas and his followers upon the Rio Grande would be of short duration. I did not mean to say by this that our difficulties with Mexico and her embittered factions were at an end, for I believe there never will be any permanent and lasting peace between that country and the United States, until she is in the hands of some other power or becomes subject to our government in some way. I simply meant by saying that all would soon be quiet on the Rio Grande, that Cortinas and all outlaws in that quarter would be driven out of Texas, and that the state would take care of her own frontier in a way that people would be safe anywhere in the settled portions of its limits.

Statesmen may dream and ponder, and they may exhaust their skill in schemes and reconciliation; treaties may be made and patched up, and confidence may be for a long time somewhat restored; but all will be of no avail. Mexico's bright days, if she ever had any, are gone. Her peaceful sun hath sunk never to rise again. A brighter sun than she ever had must transform her into a different sphere.

Seeing the recent accounts in regard to the contentions and hopeless condition of Mexico, I am induced to add a few lines to what I have already said upon the subject.

I will now suppose that the Canadas on the north of the United States, instead of being possessed as they are by a powerful and progressive people, were in the hands of a comparatively ignorant, weak and retrograding people, with a tottering government and the elements of self-destruction within it. I'll also suppose that this country, instead of being a snowy, freezing one was one of a mild and pleasant climate, and of a broad mesquite surface, covered with horses, mules, sheep, etc., abounding in the richest of mines of almost every mineral. In consequence of this weakness and inability to take care of themselves, they are taken into the United States, or under her protection in some way, and their country opened to a reciprocity of trade with her people, as are the Canadas at present, and everybody allowed to go into and trade as best they could, on an equal footing with any and all.

I say, supposing that this were the situation of the Canadas, how long would it be before the Yankees on their borders or near to them would be among them and have all the trade of the country in their hands? And how long would it be before these rich cultivating and mesquite lands, these cattle, horses, etc., would contribute to the wealth of these Yankee merchants and speculators?

Happen what will, the people of Mexico are probably to remain in their country until their race shall have become extinct or lose its identity, and they are to be consumers, and they are to labor in the cultivation of their more productive sections of country. And the Yankees, being superior merchants and tradesmen, farmers and speculators, to the Mexicans, should they settle in Mexico—which surely they will—would soon reap the profits of their trade and make "ten strikes" at the purchase of their property and at the cultivation of their lands with the labor of the peons and otherwise.

I may be considered a little too fast about this, but it seems to me to require but little penetration to see about where this Mexican people and country are going, provided the United States remain a unit and continue to flourish. There are broad daylight examples before us to tell us how this thing will probably work, although it will take many ages to put Mexico where it would be a glorious thing, could she be today.

Chance

There are no doubt many who would wonder that the lands of Texas are so cheap and so lavishingly bestowed upon capitalists, corporations, etc. There is so much of them that if, by giving a large portion of them away, capitalists could be induced to invest largely in manufacturing and railroads could be extensively constructed. The balance of her lands would then be worth much more to the state than they all now are or ever would be without such encouragement. It is only from accident that Texas has so long escaped the attention of the business and intelligent world. Had her superior adaptation to sheep and wool growing been known, and could she have been occupied by a different people when Australia was found to be so well adapted to this business, there would no doubt now be at least thirty million sheep within her limits, and other stock and business in proportion, and many of the great chances now offering to the world in Texas would have been improved long ago.

I will here state that the winter of 1859-60 has been the most severe one ever known in Texas, and the severity of this winter will no doubt be much talked of. But if such a winter is only known once in a long lifetime, and the stock upon our prairies generally live through it without the least attention and with but little loss, no one will be deterred from coming to Texas on account of such an unusual winter.

The death of bare-bellied Mexican sheep should not deter people from coming to western Texas, as they will die in an unusual cold snap, where improved sheep and other stock would hardly feel the cold.

This winter will no doubt induce many stock growers to cut a stock of hay to feed in case of emergency hereafter. It is well enough for stock growers in any part of Texas to save a stack of prairie hay to feed any animals that stay immediately about home where the grass may be short, although it might not be needed except for oxen, milch cows, work horses or something of the kind, one year in ten.

In writing a book of this kind, there are many, no doubt, who would say nothing but nice things about the country of which they might be writing. But I have considered the dark with the bright side of my subject, believing that no man of sense would expect to read of a new country that could be honestly considered as without an objection. Men are raised differently, and are of different tastes, consequently no new country would be appreciated or would suit everybody at first. But after living in western Texas a while, there are few who would leave it to live in any other country.

In conclusion, I will say that I have tried to treat my subject fairly, and that I have no interest in deceiving people in regard to Texas, neither have I any disposition to do so. I have written simply because I believe western Texas, all things considered, is the most delightful country I ever knew, and that there are tens and hundreds of thousands of people who would be much happier here than where they are. And, furthermore, I believe a large emigration of the Anglo-Saxon or American race to this part of the world would be of incalculable advantage to the country, particularly to the future of degraded Mexico.

CPSIA information can be obtained at www.ICGtesting.com
Printed in the USA
LVOW07*1324131115

461391LV00004BB/9/P

9 780988 435766